DE-Vol. 63

VIBRATIONS OF MECHANICAL SYSTEMS AND THE HISTORY OF MECHANICAL DESIGN

presented at
THE 1993 ASME DESIGN TECHNICAL CONFERENCES –
14TH BIENNIAL CONFERENCE ON MECHANICAL VIBRATION AND NOISE
ALBUQUERQUE, NEW MEXICO
SEPTEMBER 19–22, 1993

sponsored by
THE DESIGN ENGINEERING DIVISION, ASME

edited by
RAGHU ECHEMPATI
MICHIGAN TECH UNIVERSITY

JON ROGERS
SANDIA NATIONAL LABORATORIES

DENNIS MORRISON
NEW MEXICO ENGINEERING RESEARCH INSTITUTE

MING-LIANG WANG
UNIVERSITY OF NEW MEXICO

THE AMERICAN SOCIETY OF MECHANICAL ENGINEERS
345 East 47th Street ☐ United Engineering Center ☐ New York, N.Y. 10017

Statement from By-Laws: The Society shall not be responsible for statements or opinions advanced in papers . . . or printed in its publications (7.1.3)

ISBN No. 0-7918-1179-4

Library of Congress Catalog Number 93-72639

FOREWORD

This volume contains papers that were presented at the 14th Biennial ASME Conference on Vibration and Noise. The objective of the conference, as always, has been to provide a forum in which engineers and scientists could present results of their research and exchange ideas in the multi-faceted areas of Vibration and Noise.

The program chairs and editors express their appreciation to the authors for taking time to prepare the papers for this conference. The program chairs would like to express their appreciation to the members of the Conference Steering Committee, Fred Ju, Dan Segalman, and Mo Shahinpoor, for their help in developing the conference program, and to the cluster and session editors for their work in organizing the conference sessions. Finally, all organizing participants wish to express their thanks and appreciation to the ASME technical publishing staff for their assistance in supervising the production of this volume.

Raghu Echempati
Michigan Tech University

Dennis Morrison
New Mexico Engineering Research Institute

Jon Rogers
Sandia National Laboratories

Ming-Liang Wang
University of New Mexico

CONTENTS

DE-Vol. 63, Vibrations of Mechanical Systems
and the History of Mechanical Design
ASME 1993

THE ORIGINS OF ENGINEERING DESIGN

Andrew D. Dimarogonas
Department of Mechanical Engineering
Washington University
Saint Louis, Missouri

ABSTRACT

Engineering is distinguished from craft or invention by systematic development and use of intelligence and scientific knowledge. Elements of engineering design can be found in the great Potamic civilizations but systematic engineering design activity started in the ancient Greek and Hellenistic world and matured under the Romans. The renaissance and the industrial revolution revived Engineering and modern engineering design was eventually defined during the 19th Century.

INTRODUCTION

The term *Engineering* has been used, especially in literature on the *History of Engineering*, as synonymous with *Technology* and, in many instances, with *Craft* (Landels 1978, Hill 1984). What is Engineering? The famous jazz musician Louis Armstrong was once asked to define Jazz. He answered, "*if you got to ask, you 'll never know*". Schopenhauer (1911) related Engineering to the "*battle between weight and rigidity*". Aristoteles (Aristotle)[1] gave Engineering a sense of wonder: *Nature works against man's needs, because it always takes its own course. Thus, when it is necessary to do something that goes beyond Nature, the difficulties can be overcome with the assistance of Engineering. Mechanics is the name of the Engineering discipline that helps us over those difficulties; as the poet Antiphon put it, "Engineering brings the victory that Nature impedes." Technology* encompasses Craft, Invention and Engineering. There is no clear dividing line, each one of the three includes some part of the other two, as shown in Figure 1, but there are many attributes which *causa sine qua non* distinguish Engineering from Craft and Invention.

Reuleaux (1875) suggested that the earliest machine was the twirling stick used for starting fire and discussed further other early machinery such as water mills. The lever and the wedge also stem from our heritage of technology from the Paleolithic era. By the beginning of the historical age, there were substantial technological developments arrived at by craftsmen on the basis of repetition and improvement (evolution) or by inventors (revolution). Engineering is distinguished from these processes in that it arrives at a technological solution to a stated problem in a systematic manner (intelligence), see Figure 1. There is overlap and the distinguishing line cannot always be defined precisely. Numerous examples illustrating these points will be discussed in the following.

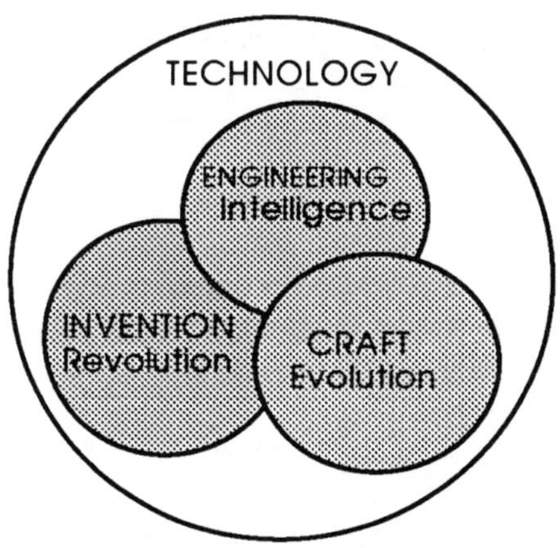

Figure 1. Technology: The Interactions of Engineering, Craft, Invention

[1] The original Greek spelling is used for Greek names. The Latin spelling is given in parentheses in the first occurence, for clarity.

ENGINEERING IN MYTHOLOGY-PROMETHEUS

In the beginning there was Chaos. The world was an arena where Gods were fighting among themselves and with the Titans, who were their uncles. Hesiodos (*circa* 700 B.C.) in his *Theogonia* (Gods' Creation) expressed the oligarchic view that Prometheus, a Titan, taught the people technical skills, and thus was responsible for their misery. Zeus punished him for this crime. In the realm of Athenian democracy, this myth took a different form. Aeschylos, in his *Prometheus Bound*, tells us the other side of the story. This tragedy has a powerful symbolism: Prometheus taught the people the alphabet, science, technology and the use of the fire. With this knowledge, they could tame nature and achieve civilization. This brings Prometheus, the symbol of Democracy, on to a collision course with Zeus, who symbolizes Oligarchy and Tyranny. Zeus orders the State to arrest Prometheus and nail him to a rock on Mount Caucasus, see Figure 2. *Prometheus Bound,* the *magnus opus* product of Aeschylos in his age of maturity (it was produced in 458 B.C. shortly before his death in 456 in Sicily) is a hymn to democracy and Engineering:

STATE:
He should pay for such a crime
and be taught a lesson, not to sap the strength of Zeus,
and to abandon his philanthropic habits *(8-11)*
OCEANIDS:
Show us and tell us all;
what Zeus found wrong in you
and why tortures you so viciously and bitterly?
PROMETHEUS:
...but I, alone, would dare and the people
were saved from going down to Hell in ashes.
...It was my decision to help the people, and I am guilty,
I do not deny it; I am now suffering great pain
and misfortunes...
...People were watching, but not seeing
people were hearing but not listening,
but as coming out of a dream all their life
they were confused and
they could not build houses full of sun
nor could they work the wood, but in dark caves
were creeping around like miserable ants.
They could not see the signs of winter
nor the blooming of spring, nor summer
fruitful with crops, but were living
without intelligence, until I showed them that the stars
regularly rise and fall.
I taught them numbers, the greatest wisdom,
and the clusters of letters,
workers of the Memory, the mother of the Muses.
And I first tamed the beasts under the yoke
and relieve men from the heaviest labor.
And I harnessed horses to the chariot,
pride of the nobility and the wealthy;
and I alone gave the sailors
ships with linen sails.
But unfortunate me! while I found such
wise crafts for the people, I have nothing
to save myself from my misfortunes *(193-267)*

The parallelism with the crucifixion of Jesus in the Christian tradition is apparent. Prometheus is the first engineer, mythical but also very real. The myth of Prometheus gives Engineering a political dimension: it is the instrument of democracy, with Engineering people can have the power to control the forces of nature and can be masters of their own destiny. This domination finds its archetype in the myth of Daedalos: He creates the labyrinth, in order to imprison the evil forces of nature and then builds himself wings with which he could defeat gravity. Paul Valery (1960), in his famous dialogue between Socrates and Phaedron, gave Phaedron these words about the engineer Eupalinos *...how marvelous, when he spoke to workmen! ... He just gave them orders and numbers...* Socrates replied: *God does just that.*

THE FIRST DESIGN ENGINEERS

One can find technological elements in the most ancient and primitive societies, i.e. in finding and storing drinkable water, in irrigation, and in tool making. Between the years 6000 and 4000 B.C., cities were built and agriculture, pottery, textiles, and metal production and processing were introduced. Thus, in the Great Empires of Mesopotamia and Egypt and the feudal societies of the East, India and China of the Potamic (great river) age, there was parallel development of crafts and technology without the concurrent development of sciences, which is the *sine qua non* of Engineering. Technical advances were arrived at by long evolution or invention and not by a conscious search for the solution of a societal problem. Moreover, the political and social systems did not allow for the liberal thinking, that is essential to the development of scientific thought and knowledge was confined to the clergy or to the ruling cast. This did not prevent the construction of monumental works in these societies, for example, the Great Pyramids and other monuments of Egypt. These were gigantic in size and complexity and must have demanded a substantial planning effort. It seems, however, that technology was developing by trial-and-error rather than by rational Engineering in the above sense. King Neter-khet's pyramid Saqqara (*circa* 2800 *B.C.*), for example, the first one to be built of stone, was originally built as a tomb (see part a in Figure 3). Imhotep, the first master builder whose name is known, later increased the dimensions of the structure in several stages, **b** to **e** (Strandh 1979). Strandh calls Imhotep "the first engineer whose name is known" but there is no evidence that

Figure 2. Prometheus torture

Imhotep fits the description of an engineer, as proposed above. In the Bible, on the other hand, the literary and historical value of which has been greatly underestimated, King Solomon (Kings, I,6,7) built a magnificent house for his Lord "... *of stone made ready before it was brought thither: so that there was neither hammer nor axe nor any tool of iron heard in the house, while it was building*". Such an undertaking would require detailed engineering planning and drawings. Even if only the symbolic character of the biblical scripts is accepted, one has to accept that in king Solomon's time the detailed engineering planning was taking place.

Another development in Egypt which could fit the description of Engineering was recorded by Herodotos (Herodotus). King Sesostris (or Rameses II, *circa* 1300 *B.C.*) distributed land among all Egyptians in rectangular plots on which he levied an annual tax. The flooding Nile could wipe out part of a plot or its boundaries and surveyors would have to remeasure the boundaries after the waters receded, so that taxes could be deducted in proportion with the amount of land lost. This need prompted the surveyors to develop quickly the necessary methodology, thought to be the foundation of Geometry (Heath 1921, 1981). However, by "Geometry" Herodotos means "Land Surveying", which was the original meaning of the term in Greek (Γαια = Earth, μετρω = measure), and not Geometry, the branch of mathematics developed by the great geometers from Thales of Miletos to Eucledes (Euclid). Moreover, some 20th Century discoveries, notably the ones reported by Neugebauer (1934) and Needham and Wang (1959), showed that these methods were known before to other Potamic civilizations, such as in Mesopotamia, India and China. People in those regions had been developing technology, sometimes in a systematic way, to solve problems in their every-day life since the Paleolithic times. The rules they had developed and used were generally of an empirical nature and no attempt had been made to derive these rules from fundamental principles in a rigorous way.

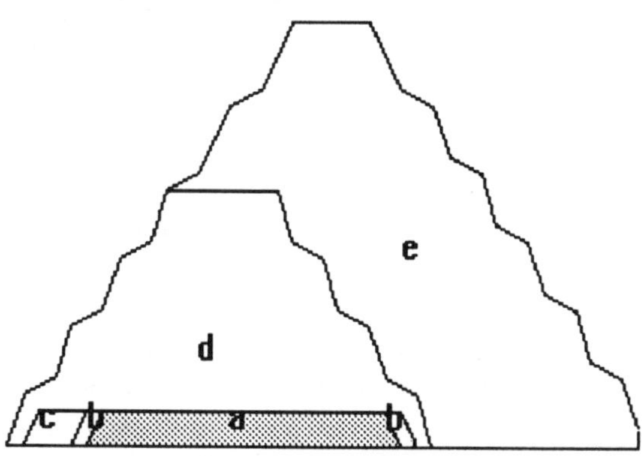

Figure 3: Construction of King Neter-khet's pyramid Saqqara by Imhotep (circa 2800 B.C.)

In ancient times, the water supply was an important problem for both city dwellers and for farmers. Naturally this need led to significant engineering developments. Herodotos tells the story of a 3x3 m in cross-section and about 800 m long tunnel which the engineer Eupalinos from Megara (near Athens) constructed on the Island of Samos[2]. Such tunnels were in use at that time they were usually dug from both ends to expedite the work. The joining of the two branches was achieved by trial and error, the two crews relying on the sound of digging to aim towards each other. If they missed, the direction of digging would be altered until the two branches met. Eupalinos used precise geometric methods to guide the excavations on the Samos Island and the two branches met almost precisely on the first try, the error was about 1 m. This, as also the Hezekiah aqueduct which was discovered recently (Keller 1974, 1982) and will be discussed in the next section, are examples of the difference between *Craft* and *Engineering*. A similar project in 152 A.D. in the Roman town of Saldae in North Africa was reported have been saved from failure by Nonius Datus, a surveyor, who was summoned after the two crews had gone the full length of the tunnel without meeting. Datus used more precise surveying methods and the project was completed (Landels 1978).

Military works were always a driving force for design innovation. Herodotos (Z31) describes the construction of two bridges over the Hellesponte, to bridge Europe and Asia so that the Persian army of King Xerxes could move forward to conquer Greece. Herodotos discusses in some detail the development of the concept and its implementation. The builders were Egyptians

[2]Herodotos, History, Γ60. The tunnel was found in the 1890 excavations.

Figure 4. Pythagoras in his research laboratory

and Phoenicians, which means that the work was considered to require not the layman but people of with considerable skill and previous experience in the art of bridge construction. First they tried a rope bridge. The Egyptians used papyrus rope while the Phoenicians used linen rope, showing that the designers were operating on long experience in a certain technique rather than any understanding of loads and material properties. Indeed, their experience was with river bridges as both of the bridges built for Xerxes were washed away by a storm soon after the work was completed. The designers paid for these failures with their lives.

Subsequently, new designers were engaged, apparently more knowledgeable with the sea, its wave action and storms. They built bridges of ships tied together, 360 ships for the north bridge and 340 for the south one. The ships were tied together with several ropes of linen and of papyrus, a deliberate redundancy to ensure against any future material failure. The surface of the bridge was paved with wood and soil and provided with side rails for safety. These bridges proved successful and Xerxes, with 1.7 million soldiers[3], was able to invade Greece.

This story of Herodotos tells us a great deal about the way major construction projects were designed in antiquity. The design was based on experience and the method was trial-and-error. Extrapolation from previous designs was very limited because the designers did not understand the physical principles behind the failure mechanisms and the scientific foundations of Engineering were yet to be developed.

ENGINEERING IN THE BIBLE

The Bible, beyond its religious and philosophical dimension, has an invaluable historical and literary utility. Although its descriptions are sketchy, because the aim was not the recording of history, there is substantial information in the script about the state of Science and Engineering in the last millennia B.C. For example, Isaiah knew around 550 B.C. that the earth is round (Isaiah 40.22).

The destruction of the Jericho walls by inducing resonance might be only a symbolic story but it informs us that in Josua's time (1200 B.C.) they knew about natural vibration, resonance and dynamics of structures. Later, king of Judah Uzziah (reigned in 825 - 745 B.C.) constructed engines of war (*invented by cunning men*) and other military works (Chronicles, II, 26.15). The careful planning in the Biblical description and the fact that Uzziah built original engines might qualify him to be the first engineer on record and the predecessor of Archimedes.

King Solomon (Kings, I,6,7) built his magnificent Temple around 950 B.C. using detailed planning and possibly detailed drawings (more discussion on this point later). There is archaeological evidence that Solomon's time marks a substantial change in the design of public buildings and other construction in Judea and Israel. It is apparent that there was substantial design engineering activity in king Solomon's court. A magnificent design plan for Jerusalem can be found in the Revelation (21,10). Of course, the text is prophetic but it shows that John new what takes to design a magnificent city.

Around 700 B.C., king Hezekiah built a tunnel to feed the waters of the Fountain of Gihon into the Siloam Pool. From the inscription found in the tunnel, one can deduct that accurate land surveying methods must have been employed.

Unfortunately, there is little additional information for the Biblical lands beyond the Biblical Scripts and it is difficult to distinguish the engineering content in the substantial technological innovations we find in the Bible.

THE BIRTH OF NATURAL SCIENCE

The wealth of empirical knowledge generated in the great Potamic civilizations of the Neolithic and bronze ages was transformed by the ancient Greeks into science (Farrington 1944). Understanding Greek science is not possible without understanding the shortcomings of Lord Bacon's theory[4] that "*technological discoveries...are much older than philosophy and the natural sciences. To say the truth, when the theory and dogmatism made their appearance, then came stagnation in the*

[3] This number of Herodotos has been disputed. Philologists and military experts estimated it to be between 60,000 and 200,000.

[4] New Organon., I, XXXV

applied fields". This is true if technology is thought to be only invention or the construction of large edifices that impress the layman. Engineering, however, was born out of the development of the natural sciences.

The general use of steel in agriculture and war, the popularization of the alphabet, and the general use of papyrus paper for book writing, were among the reasons for the rapid advancement of learning and science in ancient Greece. Moreover, the colonization of the shores of the east Mediterranean created small but dynamic and enterprising Greek communities with little or no dependence on the Metropolis. This helped the development of science in two ways: It allowed the exchange of ideas and information with the eastern civilizations and consequently the development of new ideas. Moreover, to quote Goethe, "*Human beigns can accomplish remarkable things by combining their different skills and talents; but they can only achieve something unique, something unexpected, if all their abilities come together in perfect harmony*". The most important contribution of the pioneers of Greek (and Western) science, who were the philosophers of the Ionian School of natural philosophy, was the idea that "*Nature can be explained*" (Schrödinger 1954).

The scientific method of dealing with nature started with the Ionian School of Natural Philosophy whose distinguished leader was Thales of Miletos (640-546 *B.C.*), the first of the Seven Wise Men of antiquity. He is perhaps best known for his legendary discovery of the electrical properties of yellow amber (electron) and for introducing the term *electricity*, the phenomenon of static charge that results when electron is rubbed against a wool cloth. More important, Thales introduced the concept of the logical proof for abstract propositions. Thales, who was also a very successful entrepreneur, traveled extensively in Mesopotamia and Egypt and became familiar with the knowledge of geometry (land surveying) and astronomy in those regions.

It was among the Eleatic philosophers that we can find the significant roots of logic as developed by Platon (Plato) and Aristoteles into a science which then served as an instrument for the parallel development of the natural sciences, especially mathematics and physics, by such pioneers as Pythagoras, Aristoteles, Eucledes and Archimedes.

The search for Reason led to the development of a generalized science as distinct from a set of unrelated empirical rules. The Pythagoreans, for example, sought the principles of geometry in ultimate ideas and investigated its theorems in an abstract and rigorous way (Proclos Diadochos, 410-485 A.D.). The rigorous proof, based in deductive logic and mathematical symbolism, was introduced. Experimentation was established as a method to support and lead scientific reasoning. Notable examples are the experimental investigations of sound and vibration by Pythagoras in the 6th century *B.C.* (Theon of Smyrna *circa* 150 A.D.), see Figure 4, and the experimental proof of the corporeality of air by Straton of Lampsakos (*circa* 280 *B.C.*), known to Anaxagoras since the 5th century *B.C*.

Pythagoras of Samos (*circa* 570-497 *B.C.*) was nearly a contemporary of Buddha, Confucius and Lao-Tse and can be considered a student of the Ionian School. After traveling to Babylon and Egypt, and probably India, he moved from Samos to the Croton of southern Italy and established the Pythagorean School, the first institution of higher education and scientific research. The primary contributions of the Pythagorean School were the theory of numbers and the theory of music and harmony. Pythagoras also established a rational method of measuring sound frequencies, at least integer fractions and multiples of basic sounds of musical instruments (Dimarogonas 1990), where the frequency for calibration would be the perception of trained musicians. Their ears were undoubtedly accurate enough for the needs of the time. It can be said that Pythagoras founded not only the science of acoustics but also the theory of vibration, since early enough the relation of sound and vibration was well known in his school.

Pythagoras conducted several experiments with hammers, strings, pipes and shells. He established a vibration research laboratory, the first known man-made research laboratory (Dimarogonas, 1990). Moreover, he invented the monochord, a purely scientific instrument to conduct experimental research on the vibrations of taut strings and to set a standard for vibration measurements. Furthermore, he provided experimental proof that natural frequency is a system property and does not depend on the excitation.

Contrary to the traditional belief that Galileo Galilei first observed the isochronism of the pendulum, they probably were both known long before to ancient cultures, such as the Greek and Chinese. Legend suggests that Daedalos (middle of the 2nd millennium *B.C.*) was the inventor of the pendulum. Pendulum appeared as a bob for spinning and a leveling device in paintings on vases of the 6th century *B.C.* Spinning was also known in India as early as 2500 *B.C.* The first indication of the use of the pendulum as a timing device is mentioned in Aristophanes (450-388 *B.C.*). In his *Frogs* (Ranae) he made a direct reference: *'...the music should be balanced with an oscillator'*. It can be assumed that the pendulum and, in fact, the balance which is a compound pendulum were known in 4,500 *B.C.* The dimensions of balances found or read in ancient drawings indicate that they would oscillate at about 1 Hz which is a practical frequency for a balance. It can be further assumed that they were so designed and that the isochronism was understood.

Developments in natural philosophy and the scientific method in the 6th and 5th centuries B.C., culminated in a rapid development of Engineering in the 4th to 1^{rst} centuries B.C. in the Greek and Hellenistic world to reach maturity in the Roman Empire between the first century B.C. and the 2nd century A.D.

THE ORIGINS OF SYSTEMATIC DESIGN

The first designers were the master builders of the Potamic Civilizations (India, Mesopotamia, Egypt, China). Those designers rose to the level of Engineering in the Thalassic (great seas) societies of ancient Greece and Rome. They did not know much about marketing but they knew the users of their products. They had continuous feedback concerning their acceptance and performance. During the manufacturing phase they kept in touch with the prospective users and they were always in line with the contemporary ideas about acceptable appearance and utility. Moreover, conception, design and manufacture were the work of a single person. Consequently, these products were simple and of human proportions. Much later, when mass production caused the breaking of this process into distinct smaller ones and led to the separation of design from manufacturing. However, the study of the principles underlying design activity were investigated very early in history. The first design theory was part of aesthetics where beautiful (*venustas*) meant a concept that included also functional (useful) and ethical (the good) attributes. Function and ethics were inseparable from the form. This society simply could not afford spending resources only for aesthetic pleasure. It was able however to afford a pleasing appearance for the useful goods of everyday life and to pay attention to more general societal needs.

Rapid advancement in natural sciences was followed by systematic attempts to organize knowledge in Engineering and, in particular, in Mechanical Design, developing this body of knowledge beyond the level of a mere craft.

Kinematics and Mechanism design have a distinct place in the history of Engineering because they comprised the very first of its divisions that received a mathematical foundation. Heracletos (Heraclitus) of Ephesos (*circa* 550-475) appears to be the first that separated the study of motion itself from dynamics, the forces causing the motion, and introduced the principle of *retribution*, or *change*, in the motion of celestial bodies.

The first known written record of the word *machine* appears in Homeros (Homer) (*Iliad*, 7th century *B.C.)* and Herodotos (*Histories*, 5th century B.C.) to describe political manipulation!. It was not used with its modern meaning until Aeschylos (Aeschylus) used the term to describe the theatrical device used to bring the gods or the heroes of the drama on stage; whence the Latin term *Deus ex machina. Mechanema-Mechanism* in turn, as used by Aristophanes, means *an assemblage of machines.* None of these theatrical machines, made of perishable materials, is extant. However, there are numerous references to such machines in extant Greek plays and also in vase paintings from which they can be reconstructed: They were large mechanisms consisting of booms, wheels and ropes that could raise weights perhaps as great as one ton and, in some cases, move them back-and-forth violently to depict space travel, when the play demanded it.

The designers and builders of these mechanisms were called by Aristophanes *mechanopoioi* (machine makers), meaning *mechanical engineers* in the modern terminology and identical with the corresponding German term *Maschinenbau Ingenieurs.* They designed and built the M*echane* not by evolution nor by invention but to the order and specifications of the playwright, Aeschylos in this case, and they truly deserve to be called the first *Mechanical Designers.*

Very few details are known about the *Mechanes* but it is certain that they were substantial mechanisms for path and motion generation. They were probably several meters high and operated above the stage and roof of the theater. In a fully retracted position a *Mechane* was hidden behind the stage, so it would have to have more than one moving link and it could curry substantial load. In Aeschylos' "Prometheus Bound", at the entrances of the Oceanids and Oceanus, a 15 member chorus plus chariot and horses are lifted in the air, a total estimated weight of one ton. The simultaneous entrance of Oceanos must have necessitated the use of a second *Mechane* or else a very complicated single one. *Mechane* could also be used to create very violent motions. In the *Peace* of Aristophanes, Trygaeus straddles a huge dung-beetle and flies upward to the abode of Zeus in a swinging, violent motion to convey a sense of space flying, while he chants eighteen lines of anapest dimeters; then he is lowered back to the terrace from which he started. As he flies upwards, he asks to the stage engineer to be careful.

The M*echane* must have been well-balanced. Antiphanes in *Athenaeus* complains that when tragic poets have nothing left to say, they "raise the *Mechane* like a finger". Such undertakings would require mechanisms enormous in both size and complexity. The technical challenge is similar to that in certain scenes of opera, such as the flying maidens in *Valkyrie*, the swans in *Lohengrin* and the dragon in *Siegfried.*

The principles of statics and dynamics were discussed by Aristoteles in *Problems of Machines (MHXANIKA,* for the translation see below), the first extant treatise of this kind, written probably by one of Aristoteles' students in Lyceum. *MHXANIKA* starts with the definition of *Machine,* which in that era was synonymous with *Mechanism.* In fact, mechanisms were the only machines known. To quote Aristoteles, "*Machine is a means of altering nature's course".* The author further discusses several purely kinematic aspects of mechanisms, such as:

a. the vectorial character of velocity, the superposition of velocities and the parallelogram law for velocity addition;

b. the concepts of absolute and relative velocity of points along a link of a machine;

c. circular motion: the velocity of every point on a wheel rotating about its geometric center is directed along a tangent to its circular path and is proportional to the radius of the circle and the angular velocity;

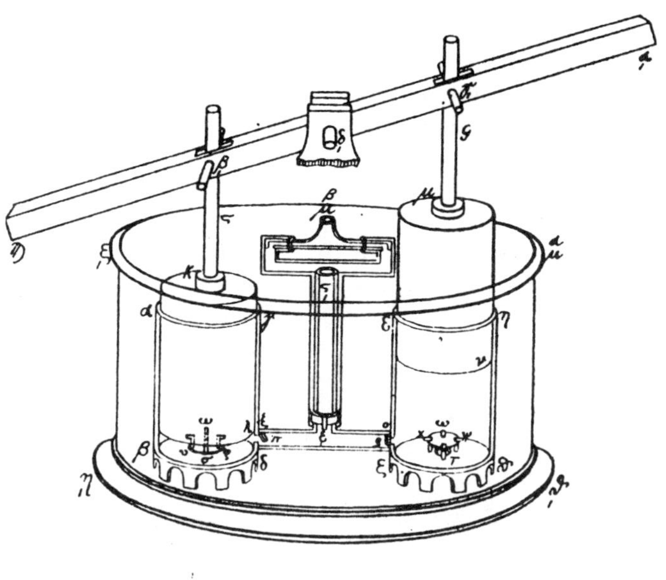

Figure 5. Heron's double-piston water pump

d. the motion of two wheels rolling against one another without slipping; they rotate in opposite directions;

e. the rhomboid four-bar linkage and the relative velocities of the opposing joints.

For all of these aspects above, Aristoteles developed rational geometric methods and proofs. Moreover, *MHXANIKA* contains remarkable discussions of the mechanics of the lever, the balance, the wedge, rolling friction, the strength of beams, impact, mechanical advantage, and the difference between static and kinematic friction. The second successor of Aristoteles in the Lyceum was Straton of Lampsakos (*circa* 270 *B.C.*) who introduced an important kinematic criterion of equilibrium, the Principle of Virtual Velocities.

The decline of Greece-proper coincided with the rise of Alexandria, founded in honor of Alexander the Great (356-323 *B.C.*) in the Nile Delta. This city became the most important scientific center in the world. In its University, the *Museum* (meaning, the house of Muses, who were the protectresses of the Arts and Sciences) flourished a number of great mathematicians and engineers. One of the most important engineering professors of the Museum was Ktesibios (*circa* 283-247 *B.C.*), the inventor of the precision water clock. He left a lot of writings which were subsequently lost and only references to his writings by his students, notably Philon and Heron, are extant.

Purely mechanical treatises on machinery go back to the 4th century *B.C.* One of Platon's contemporaries and friends and a student of Pythagoras, Archytas Tarantinos (*circa* 400-365 *B.C.*), is said to have written the first systematic treatise on machines based on mathematical principles. This, too, is lost. It is known, however, that Archytas built an air-propelled, flying wooden dove (Aulus Gellius *circa* 150 A.D.). Details about Archytas dove are not known but it is the first flying machine.

There exist also some fragments and references to an extensive treatise, *Machine Syntaxis* (synonymous to *Synthesis*), by Philon of Byzantion (*circa* 250 *B.C.*). He was a student of Ktesibios at the Museum and his treatise dealt among other things with the idea of a small number of simple elements which constitute every machine. Different machines are constituted from different syntheses of these basic elements. Most of this treatise was also lost and only parts of it as well as references in other works are extant. However, a comprehensive mechanical treatise of Heron (Hero) of Alexandria (*circa* first Century A.D.)[5] is extant in an Arab translation. Its title *MHXANIKA* (synonymous with that of Aristoteles' treatise) was mistranslated and misunderstood. It is generally known as *Mechanics* (*MHXANIKH*) while the correct meaning of the title is *Problems of Machines.* The Greek word *MHXANH* (pronounced *Mechane* in Greek) means *Machine*. It should be noted that *MHXANIKA* is a not a guide for the design of mechanical devices, because Heron wrote such texts for particular categories of devices, such as pneumatic machines, automata, optical instruments, balances and artillery machines. *Problems of Machines* is really a study of design methodology. Heron separates the study of particular machines and the general concepts of machines from the study of standardized elements that, in different combinations, constitute any machine. He introduced five simple mechanical elements for the solution of the general problem of moving a weight with a given force: wheel and axle, lever, windlass, wedge, and screw. In fact, he asserts that all five solutions are physical devices embodying the lever principle, a simple function module in terms of contemporary literature. Heron's *Problems of Machines* gave the first systematic development of design solutions to a given mechanism problem, Figure 5. It is to be noted that a substantial part of Heron's work is based on earlier (lost) works of Philon and Ktesibios.

Heron was aware of the difference between the craftsman and the designer of mechanical devices. Pappos of Alexandria (4th century A.D.) expounds on this:

*"The mechanicians (*more precisely: the *mechanical engineers, OI MHXANIKOI) of Heron's School told us that the study of machines consists of theoretical and practical part. The theoretical part includes the natural sciences while the practical part consists of the engineering disciplines. They postulate that a necessary condition for an able designer*[6] *of mechanical*

[5] Heron is the most proliferant writer of his time but we know very little about his life. From his writings, he seems to predate Vitruvius but an observation of a sun's eclipse of the year 62 A.D. that he described in "Dioptra" places him in the first century A.D.

[6] Most scholars translated this word erroneously as "inventor" . Heron uses the word ευρετης = discoverer, not by chance but by thought (Lindell and

devices is a solid background in both natural sciences and practical skills".

In modern terminology, this could be a reasonable account of what should be required of a mechanical engineer today. "*The Heron School*", Pappos continues, "*recognized the need for specialization, for it is not possible for someone to have competence in all the related fields. Further, they classified the Mechanical Engineering specialties in five categories: handling of weights, military engineering, hydraulics, marvelous device making, sphere making*".

Heron's works *Automata Making, Pneumatics* and *On Water Clocks* (the latter is not extant) describe a number of devices that most of contemporary commentators consider an exercise in toy making. A careful observer can see instead that *Automata Making* and *Pneumatics* are design studies wherein Heron develops alternative solutions to some basic design problems. Heron was explicit about his intention to develop original designs: "*Furthermore, one must avoid the predecessor's designs, so that the device will appear as something new*".

The end of the Alexandrian era marked the eclipse of the ancient Greek science and with it the systematic study of the design of machines became stagnant for a long period of time. During that time, the enormous emphasis of the Romans on practicability led to the separation of the functional considerations from those of form and aesthetics.

ROMAN ENGINEERING DESIGN

As the slavery-based society of Rome reached maturity, productivity fell and once again utility prevailed over form in design considerations. Romans were not great philosophers or artists but they were great engineers and designers. They recorded and adopted the knowledge produced in the earlier dynamic societies of Greece, Egypt and Mesopotamia to accomplish great feats of civil and mechanical engineering works needed by the Empire. Aesthetic and ethical dimensions were not important. Romans were mostly unaware the Greek mathematics until the second century A.D. when Greek mathematical works started being translated into Latin. There is a substantial number of treatises in architectural and mechanical design, mainly encyclopedic in nature; the one by Vitruvius was the most notable. The Romans further gave the world sophisticated legal and administrative systems and separated the professions of Civil and Mechanical Engineering. Moreover, they were careful recorders of the design "rules of thumb". Analytical reasoning gave way to pure empiricism and improvement by repetition. Progress, sometimes very substantial, such as the invention of roller bearings, which according to some findings might have been used by Romans and Chinese in the second and first centuries B.C. (Hartenberg and Denavit 1964), seems to have been the result of chance discoveries followed by trial and error experiments rather than the sustained use of intellect combined with analysis and experiments designed to explain the physical phenomena involved. In this respect, the design of fairly advanced mechanisms such as cranes and linkages must be classified as purely empirical.

Commentators on the classics flourished in Rome. They not only preserved most of the classical culture but made substantial advances of their own. Most notable was the architect Marcus Vitruvius Polio whose ten books *On Architecture* contained important material on the history of technology and on the design of machinery. Vitruvius defined a machine as "*a combination of timbers fastened together, chiefly efficacious in moving great weights*" and his book is imbued (Benvenuto 1991) by the words *firmitas, utilitas, venustas* (strength, utility, beauty) .

ANALYTICAL DESIGN

Aristoteles in his Μηχανικα - *Mechanical Problems* introduced the analytical method for the study of machines and mechanisms (Benvenuto 1991). He posed a number of problems on the equilibrium and stability of levers and the balance, and others problems related to shipbuilding, tribology, and the wedge.

Archimedes (*circa* 287-212 *B.C.*), who was one of the Museum students but lived in Syracuse (Sicily), mastered all the disciplines of mathematics and engineering and wrote separate treatises in the above areas. Archimedes was both a great engineer and inventor but his books concentrated on applied mathematics and mechanics and rigorous mathematical proofs. There is a clear distinction between his and the later works of Heron. Even their methodologies were different: where mathematical rigor was used by Archimedes work, Heron applied numerical methods. He reported on these in his *Metrica*, a treatise of numerical methods for area computations and for the approximation of roots.

Archimedes systematized the design of simple machines and the study of their functions. He was probably the inventor of the compound pulley and developed a rigorous theory of levers and the kinematics of the screw. Plutarchos and Polybios described giant mechanisms for lifting ships from the sea, ship-burning mirrors and a steam gun designed and built by Archimedes. The latter allegedly fascinated Leonardo da Vinci. However, the validity of these stories is questionable.

Philon of Byzantium wrote a treatise on artillery *(Technology of Arrow Making)* in which he described an analytical method for the design of ballistae (heavy stone throwers, Figure 6). Such machines were used by the peoples in the East. In the Second

---

Scott 1843), and not εφευρετης = inventor.

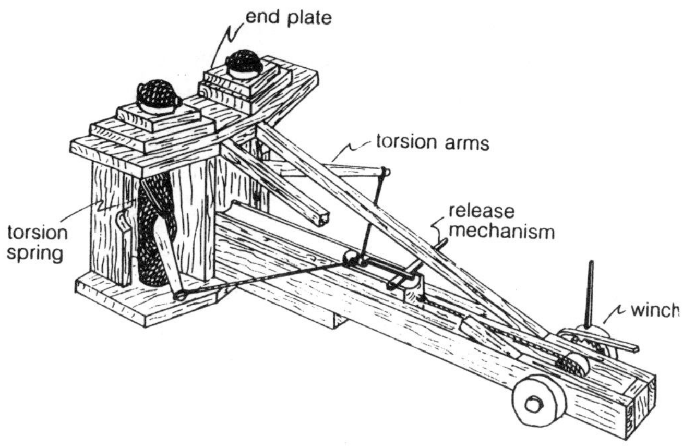

Figure 6: Balista. From (Soedel & Foley 1979)

Book of Chronicles, for example (26,15), Uzziah the King of Judah (779-740 B.C.), talks about such machines. Philon stands out because he has devised a design equation obtained mathematically on the basis of certain assumptions. In the case of the balistae Philon's equation was $d = (11/10)\sqrt[3]{100m}$, where d is the diameter of the twisted gut skeins providing the potential energy, in dactyls, and m the mass of the stone, in hundreds of drachmas. In conjunction with this formula, Philon introduced the idea of *similitude*, stating that larger machines can be designed on the basis of the dimensions of smaller ones with the formula $d_1/d_2 = (m_1/m_2)^{1/3}$. Moreover, Philon's formula for the balistae was the first developed design equation in the contemporary sense that related directly the design objective (the payload) to the design parameter (the torsional spring diameter). In Philon's formula, the cube root relationship was derived theoretically (though not correctly) while the numerical constant 100 was derived from the units used and the constant 11/10 was used as a safety margin to account for the error involved in the approximation for the cube root, introducing the safety factor. Cotterell and Kamminga (1990) proved Philon's formula in the form $d = c\sqrt[3]{m}$, where c is a constant. Marcus Vitruvius Polio (*On Architecture)* proposed a number of pairs (m, d), apparently obtained experimentally, which fit closely the equation $d = 3.3\sqrt[3]{m}+(1/60)(60-m)$, in Roman units (Cohen & Drabkin, 1958).

This method of systematic design was further exemplified by Heron in his *Arrow Making* (artillery) where he also introduced the idea of the sensitivity of the design to variations in the design parameters. It is not strange that Heron, driven by abstract reasoning and not by practical needs, designed devices using very advanced principles, such as those that can be conceived as very early models for the reaction machine (turbine) and the piston engine. Moreover, his design of machines was based on mathematical analysis and engineering science, and is thus distinguished from mere empiricism.

In his *Problems of Machines* Heron made extensive use of analytical methods for the study of machines and mechanisms. His static and kinematic analysis was based on the conservation of energy ($F_1s_1 = F_2s_2$) although he did not recognize the product Fs as being energy (or work). On this basis he studied the flow of motion and force in a machine. He also developed a method of number synthesis, for finding the number of pulleys needed to raise a weight, given the weight and the pulling force (I.1), and a methodology for gear sizing on the basis of the desired mechanical advantage and the design of the screw threads (I. 19), Figure 7.

MACHINE ELEMENTS

Most of today's machine elements, beyond the ones considered by Heron, were known by the third century *B.C.* Homer (Iliad 16.775, Odyssey 24.39) made reference to the crank, Philon referred to the connecting rod (Belopoeica 74.22),

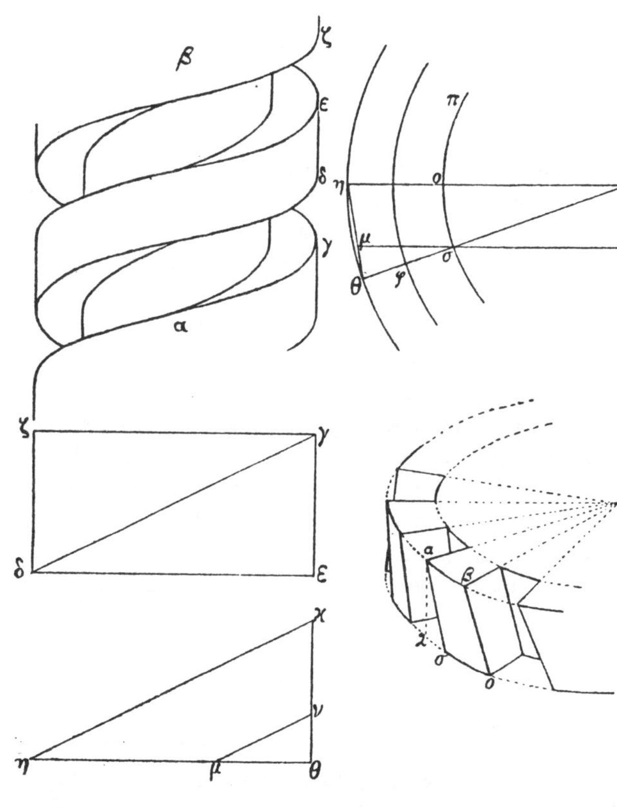

Figure 7. Heron's design of screw threads

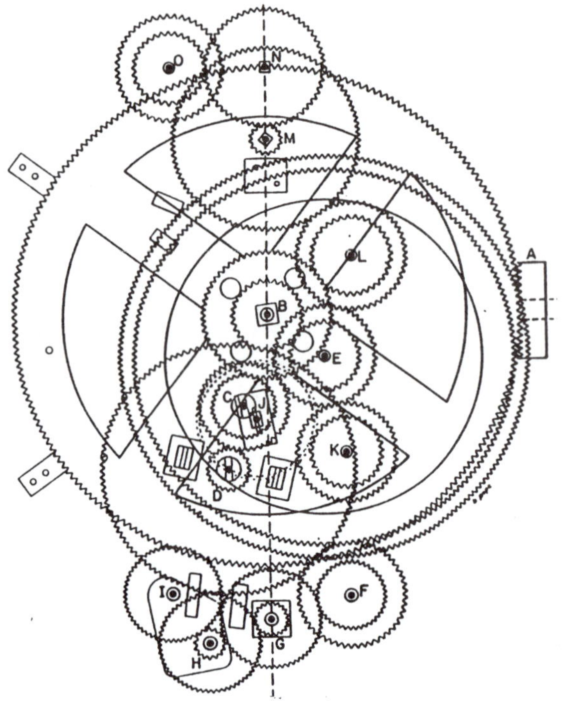

Figure 8. The Antikythera Computer

Strabon the Geographer referred to the linkage (Chrestomathiae 2.1.30) and Aristoteles posed a number of problems related to sliders (Problemata 936-15, 939-26, Problems of Machines 854-19). Crank and follower mechanisms were widely used in antiquity for many purposes, including the foot-operated lathe.

Gears were discussed in Aristoteles and were well-known to Archimedes and the Alexandrian engineers. Almost concurrently with the decline of Alexandria, the differential gear was used in China. There is evidence that at the 3rd century A.D. the Chinese built a south-pointing chariot based on the principle of the cog-wheel differential to guide merchants traveling along the Tibetan plateau from China to Samarkand and the Mediterranean. It is certain, however, that this could not been quite true due to the accumulation of error and probably there was only a model showing that the rulers had supernatural powers, similar to Heron's designs for church miracles. However, the description of the device indicated ample knowledge of the kinematics of gears. An ingenious geared mechanism for astronomical computations dating to the first century *B.C.* was found in a ship wreck off Antikythera, an island between Crete and Peloponnesos, Figure 8. Pappos (Mathematical Collection VIII, 1-5) posed a synthesis problem, to find the diameter of a gear with a known number of teeth and a given mating gear.

DESIGN FOR STRENGTH

It is generally believed that the ancients did not understand the flexure of beams and the strength of columns, two subjects which are basic to the design of engineering structures. The earlier theory of bending of beams is generally credited to Galileo. On the other hand, is must be admitted that the ancients understood strength very thoroughly, given the size and durability of ancient structures. It is true that many structures, especially buildings, were so overdesigned that they did not require fundamental knowledge of the principles of structural design but this was not always so. Moreover, some of the structures were unique, making their conception and design the work of an engineer and not a craftsman.

Aristoteles knew that the strength of a beam is inversely proportional to the bending moment (*Mechanical Problems*-14). He showed that in the simply supported beam the reactions are inversely proportional to the distance from the load and that in the cantilever beam the reaction moment is proportional to the distance of the load from the support. Beams in many monuments were tapered towards the two ends. Columns were tapered towards the upper end. An understanding of the structural behavior of the beam was demonstrated by Mnesicles who placed steel reinforcements in the beams of the Propylaia of the Acropolis of Athens.

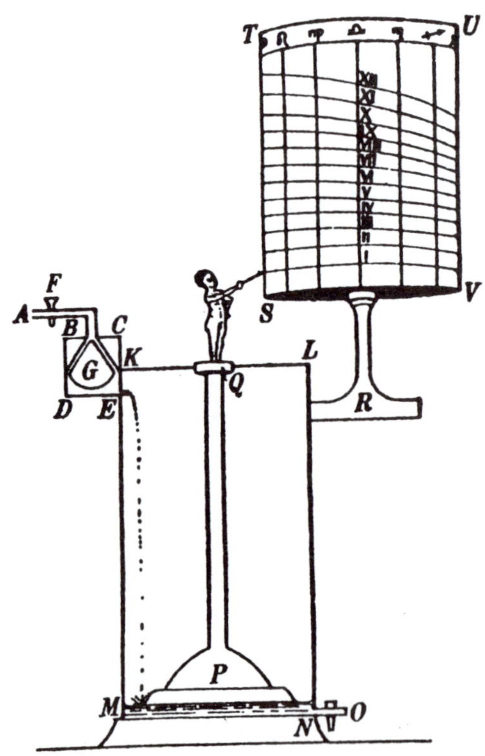

Figure 9: The Waterclock of Ktesibios.

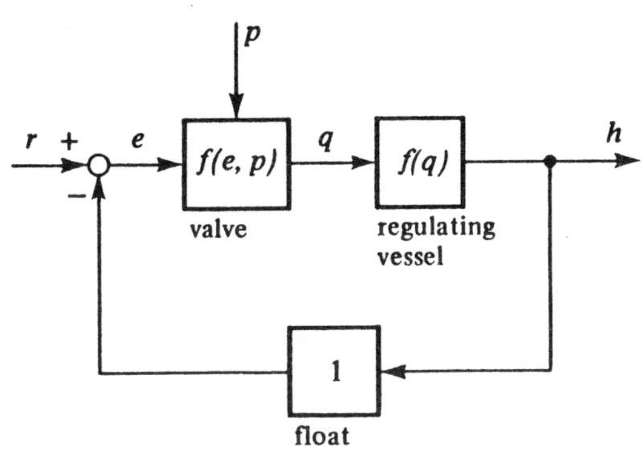

Figure 10: Block Diagram for Ktesibios Flow Regulator (Mayr 1969)

DESIGN OF SYSTEMS

The idea of the automaton was well-established in the Alexandrian times. The Archytas dove, discussed above, was not the only example. Vitruvius (1st century *B.C.*) described how Ktesibios designed a device raising a mirror in a barber shop. In fact, this can be considered the first original mechanism that was designed to order on the basis of engineering reasoning. Heron developed a great number of automata, even including vending machines. It should be noted that these devices were original designs employing advanced principles such as feedback control and not simply devices that were crafted in a gradual evolutionary process.

The oldest known design of a feedback control system is attributed to Ktesibios for the flow regulator of his water clock (Vitruvius, 1st Century B.C.). In the latter, Figure 9, a floating inverted cone G regulates the water flow so as to keep the of the water level and hence the water flow constant. Apparently Ktesibios understood the effect of pressure on flow, and also the effect of nozzle area. Moreover, he recognized the necessity for accurate measurement and he made the nozzle out a gemstone so that its internal passage could resist flow-induced errosio. A block diagram of the float Regulator of Ktesibios is shown in Figure 10, (Mayr 1970).

A substantial number of feedback control systems, mostly flow regulators, were designed by Ktesibios' students, Philon of Byzantium and later by Heron of Alexandria. They were prompted by the need for accurate water clocks. This tradition was continued in the Arab world with a great number of similar or different embodiments inspired by the three great designers of antiquity: Ktesibios, Philon, Heron.

DESIGN DRAWINGS

It is generally believed that engineering drawings based on the conventional orthographic projections is a recent development. This is not so. There are many indications that such drawings were used in the Potamic age. There is a status of a Gudea who lived about 2200 B.C. showing him seated with drafting materials and a plan on his lap, Figure 11. Most surviving drawings are sketches only. Many drawings survived in the engineering texts written by Ktesibios, Heron and Vitruvius. Since these were redrawn by scribes, several times perhaps, it is not certain whether these drawings are originals or exact copies. Recently, substantial construction drawings were discovered at the temple of Apollon at Didyma (Haselberger 1981) on the east coast of Asia Minor. The Ionian architects Daphnis of Miletos and Paionios of Effesos began planning for the temple about 334 B.C. The construction went on for 600 years but was never completed. On the surviving walls of the temple there are 200 m^2 of full-scale orthographic projections of building parts. The walls were originally covered with reddish chalk and then the drawings were inscribed with sharp instruments so that the lines would be white against a dark red background. There are signs of corrections, which could be made easily by covering the mistakes with red chalk and drawing the lines again. Top and side views of a column are shown in Figure

Figure 11: Gudea with the drawing on his lap. (Kirby *et al* 1956)

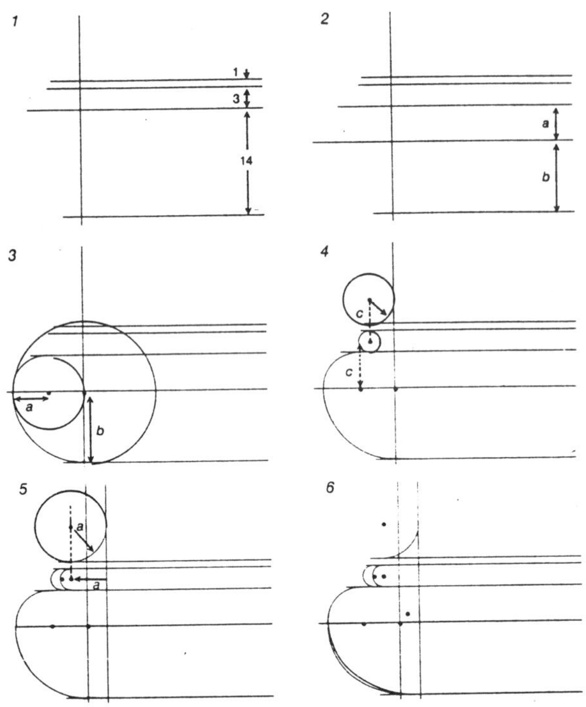

Figure 12: Orthographic projections of a column in Apollo Temple, Didyma.

12. The drawings also show the sequence of the construction of the curved lines of the base of the column. One important aspect of these drawings is the construction of the large radius of curvature lines that for aesthetic reasons (εντασις-subtle touch) replaced the long straight lines.

Since the radius of curvature could approach one kilometer, the arc could not be drawn directly. One such drawing was made in 1:16 scale, using Greek dactyls (1 dactyl = 1.85 cm) instead of feet (29.6 cm). Then, a second drawing was drawn in 1:1 scale. The 1:16 scale was very convenient because they could go back-and-forth between the scaled and the full-size drawings by purely geometric methods (doubling or halving the dimensions). Parts of the floor plan were also found, and showed that the detailed plans were made on a precise rectangular grid. There is even evidence (Dilke 1971) that Hipodamos of Miletos redesigned Piraeus and Miletus according to a master plan that was drawn on a rectangular grid.

MANUFACTURING TOLERANCES-QUALITY

Pliny the Elder in *Historia Naturalis* described in great detail specifications for the manufacturing of papyrus paper. Not only was the process standardized but also the dimensions of the sheets, in particular their width. Sheet width was important for the convenience and efficiency of the scribes. In fact, there is an on-going 50 year philological debate about the apparent omission by Pliny of the length of the sheet, in words of one author, " *Pliny's failure to mention the height of the roll*" (Johnson 1993). There was a good reason for Pliny's omission of a length specification. The quantity of pulp was measured and then spread over a surface so that the width, which was closely controlled for reasons discussed above, would be as desired. Because of an inability to control layer thickness distribution, the length of the sheet varied with the quantity of pulp being spread and the actual thickness distribution. In other words, the length varied randomly. This is illustrated in Figure 13 where the distributions of heights and widths measured on surveiving Egyptian scrolls are shown (Johnson 1993). The number of measurements is small but one can see the difference: The width histogram suggests a biased distribution while the height histogram looks more like a normal distribution.

Furthermore, Pliny also described the different standard qualities of the paper and the standard width of each, surface finish, texture and porosity.

DESIGN CODES AND STANDARDS

At a very early stage designers found that standardization is very important in design for both interchangeability and reducing manufacturing cost. Consider, for example, the standardization of the plumbing used to supply water supply in the city of Rome.

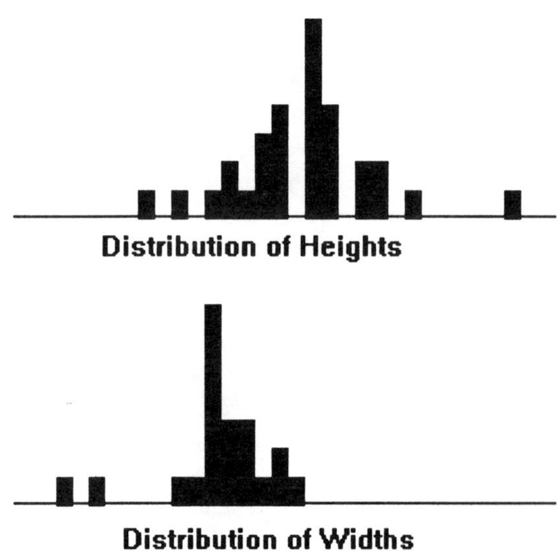

Figure 13. Distribution of dimensions of papyrus rolls.

The lead pipes had standard sizes which follow very closely the Renard Series $\sqrt[m]{n}$, n=1, 2, 3, ..., where m is the series index (Dimarogonas 1988). In this series, for example, R10 means m=10. Such series are used to standardize a small number of dimensions within a range of desirable dimensions in order to optimize the inventory of manufacturing tools and fixtures. The lead pipe diameters set by the Roman standards had sizes 5, 8, 10, 15, 20, 30, 50, 60, 100 digits (1 digit = 4.3 mm). This series is remarkably close to the R6.75 series.

ENVIRONMENTAL ISSUES IN DESIGN

There is evidence that there was concern for environmental issues in very early times. Hippocrates (5th Century B.C.) wrote a treatise *Airs, Waters and Places*, in which he discussed the problem of water pollution (Landels 1978). In the design of the Roman aqueducts, elaborate measures were taken to guard against water pollution. Vitruvius (De Architectura) advised against the use of lead pipes, to protect the public from lead poisoning that was apparent in the workers in lead-processing shops. He recognized that the lead oxide (*cerussa*) was the toxic substance responsible for lead poisoning. It is also known that noise pollution was a problem in ancient Rome and vehicles were not allowed to operate in the center of the city during certain hours.

ENGINEERING DESIGN IN THE ARAB WORLD

The Arabs played an important role in the preservation of the of the Greek science and engineering and made substantial contributions of their own. They devised ingenious mechanisms with high degree of automation and control. One of the Arabic writers of this era was ibn al-Razza al-Jazari who in his extensive and beautifully illustrated treatise on machines, *Book of Knowledge of Ingenious Mechanical Devices* (1206), described a great number of ingenious mechanisms and automata. However, he gave no indication of a design methodology although some evidence of systematic systematic development can be seen. One aspect of al-Jazari's book, which has gone largely unnoticed, is the systematic development of design concepts for particular design tasks. He divided his book into different *Categories*, each one describing a great variety of solutions to a specific problem. In category I, he described six different clocks based on the principle of water flow and four of them employing the burning of a candle. Category II considers 10 solutions to the problem of drinking. Categories III, IV and V dealt with the problems of hand washing and phlebotomy (10 solutions), fountains and perpetual flutes (10 solutions), raising water (5 solutions), respectively. Later writers who have commented on al-Jazari's work consider his designs as inventions of devices intended mostly for the entertainment of Muslim feudal lords. Many, even dismiss al-Jazari's work altogether as exercises in toy-making, but the multiplicity of solutions given in the various categories (5 or 10) clearly indicates to the knowledgeable reader that there were studies of design alternatives embodying a certain design principle. Thus, all Al-Jazari's work is really a study in systematic design!

Another, earlier book, called *The Book of Ingenious Devices* written by the brothers Banu Musa bin Shakir (ca 850 A.D.) contains a wealth of systematic design studies. It presented conceptual designs for nine pitchers, forty jars, seven flasks, and ten basins.

ENGINEERING DESIGN IN THE MIDDLE AGES

In the first half of the second millennium A.D. there was massive emigration of Greek scholars from Constantinople to Western Europe. This resulted in a substantial dissemination of the works of Aristoteles and a renewed interest in mechanics and kinematics, in particular. Jean Buridan and Albert of Rickmersdorf (both *circa* 1350), the latter known also as Albert of Saxony, were both rectors of the University of Paris at different times and were instrumental in reviving the writings of Aristoteles by publishing translations of his works and writing commentaries on his theories (Dugas, 1955).

Nicole Oresme (1330-1382), a Buridan disciple and grand master of the College of Navarre, was given by Charles V the task of translating into French some of the works of Aristoteles. Oresme wrote treatises of his own on mechanics, introduced the use of the Cartesian coordinate system and studied uniformly accelerated motion. He derived the equation $s=\gamma t^2/2$ for the distance traveled, in contemporary terminology. Moreover, he introduced the basic ideas of continuously varying quantities (Dugas 1955) thus he laid the foundations of continuum mechanics.

The diffusion of the Roman culture into the highly religious medieval Europe, shifted the emphasis on the practical needs of worshipping God. In the medieval monasteries, important mechanical devices were conceived. Again, aesthetics became secondary to pragmatic and ethical dimensions. Intellect was reserved for religious investigations and technological advancements had came as they did in Roman times. Design methodology returned to the level of a craft, according to the testimony of several authors, and no noticeable advancements were recorded until the time of Leonardo da Vinci. Although Leonardo did not explicitly enunciate a specific design methodology, in his notebooks one can see a systematic development of design concepts to arrive at specific solutions. This accomplishments had little impact on his contemporaries, however, because of Leonardo's secretive nature; his notebooks were only published centuries later. Leonardo's designs were in the style of the sixteenth century, in pictorial books with a variety of drawings of machines. They were easy to draw and also much more understandable to craftsmen than text and mathematics which indicates that the building of machinery then was just a craft.

The need for more accurate time keeping in the monasteries brought about the development of the pendulum clock, a remarkable mechanism that had a paramount effect in the development of modern occidental science and civilization starting with Galileo Galilei (1564-1642). The predecessors of the pendulum clock were the water clock, known to the ancient Assyrians and Egyptians and developed to a precision instrument by the Ktesibios School of Alexandria, and the mechanical clock, which can rightly be called *the mother of all machines*. The latter was gravity driven with an escapement control. The origin of the mechanical clock is placed as early as 1335, but it is probable that the clock mentioned by Dante (1235-1321) in his *Paradiso* was a mechanical clock. The mechanical escapement is usually attributed to the French architect Villard de Honnecourt, and shown in his *Album*, a book of drawings. The mechanical clock played a fundamental role in the revival of mechanical design.

RENAISSANCE AND THE INDUSTRIAL REVOLUTION

The early modern era is highlighted with the works of Galileo and Newton and included the early stages of mechanization and the industrial revolution. The utilization of high density chemical energy and the consequent high power per unit of machinery volume spawned numerous problems in machine dynamics. The contemporary development of calculus and continuum mechanics led to the rapid development of mechanics by the mid-19th century. By the time of Galileo and Newton, classical Physics and Mechanics were developed much further than is generally assumed. The fundamental contribution of Galileo and Newton is the revival and redefinition of these sciences at just as greater progress was being demanded of natural science. There have been several previous attempts at the revival of physics and mechanics during the first half of the second millennium A.D.. The time, however, was not yet propitious.

About the seventeenth century, as societal development moved from the Mediterranean region to western Europe, political power was concentrated in the hands of centralized, monarchical governments. The concomitant generation and concentration of capital allowed ample patronage of arts and crafts. The aim of such patronage was to elevate the grandeur of life for the nobility. Craftsmen became organized in guilds which functioned mainly to preserve their design secrets. However at the same time the guilds prevented the crafts from developing beyond mere empiricism.

By the second half of the eighteenth century, engineering design had begun to influence life beyond the royal court and the nobility. After the Revolution the royal manufactories of France entered into commercial competition and the products made for the general public began to share the high aesthetic quality previously shared by the royalty.

Developments in machines and mechanisms during the 17th and 18th centuries were stimulated mainly by the development of the steam engine just as the development of mechanical clocks provided the stimulus in the 15th and 16th centuries. Many of the advances in mechanical design in this era can be traced to some genius inventor, but many were conceived by the systematic reasoning of the engineer. Watt (1736-1819), in a letter to his son in November 1808, described the methodology he utilized to design a linkage to trace an approximately straight path. He was thus the first to consider coupler points and the systematic synthesis of linkages by means of kinematic considerations. Euler (1707-1783), a near contemporary of Watt, laid the foundations of kinematic analysis by treating planar motion as a superposition of translation of a point and rotation about it (Euler, 1736-1742) and made a clear distinction between the kinematics and dynamics of a mechanism, the former having to do solely with the motion without reference to the forces causing it. (Euler 1740) An important addition to the plane kinematics of Euler was the discovery by Johann Bernoulli of the instantaneous center of rotation in 1742.

Since ancient times, authors such as Valturius (1472), Agricola (1550), Ramelli (1580), and Branca (1629) dealt either with whole machines or with their separate elements. Leupold (1724) seemed to be the first to have considered a mechanism, i.e., a system consisting of different elements which are in some functional relationship with one another and classified by the transformations of motion which they effected. He described mechanisms that convert one type of motion to another.

DESIGN OF MACHINES AND MECHANISMS

Ecole Polytechnique, which was established in Paris in 1794, was the first school to separate the study of kinematics from the study of machinery, in general as proposed by Monge (1746-1818), the founder of descriptive geometry and one of the founders of the Ecole Polytechnique, and L.N.M Carnot (1753-1823) and elaborated by his colleague Hachette in his book *Traite elementaire des machines* (1811). The latter, presented a systematic classification of mechanisms by capability and devised a classification chart and formed the basis for the kinematics course at the Ecole Polytechnique and later at the West Point. Hachette proposed the classification of the machine elements into six *orders*: *recepteurs*, receiving the motion from prime movers, *communicateurs*, passing-on the motion, *modificateurs*, changing the type of the motion, *supports, regulateurs*, transmission ratio regulators, *operateures*, performing the machine function.

The separation of kinematics as a separate discipline was concluded by Lanz and Betancourt (1808) in their *Essai sur le composition des machines* which, with the Hachette's book are the first textbooks in kinematics. Lanz and Betancourt have separated the study of mechanisms on the basis of their motion, i.e. rectilinear, rotary, continuous, reciprocating in ten classes.

Borgnis (1818) in his *Traite complet de mechanique* further classified mechanisms into classes based on the function, such as receiver, transmitter, regulator, etc, which received further development in the works of Coriolis and Poncelet. Reuleaux disputed this approach (*Amicus Plato, sed magis amica veritas*) although he used some of these ideas in his work.

In 1830 Ampere published his *Essai sur la philosophie des sciences* defined the machine *...an instrument by the help of which the direction and velocity of a given motion can be altered.* He excluded forces from kinematic analysis and gave to the study of such machine the name *Kinematics* from the Greek word *KINHMA* = motion and the Science of Kinematics was born (*Kinetics* would be a more accurate expression, but it was used already with another meaning). In the new science *The movements are to be studied as movements, independent of the power creating them, in the same way as we observe the static bodies around us, especially in such combinations as we call machines...This science (Kinematics) ought to include all that can be said with respect to motion in its different kinds, independently of the forces by which it is produced.*

In 1831, Coriolis was investigating relative motions and posed the fundamental problem of kinematic analysis: *to find the motion of any machine in which certain parts are moved in a given way* and introduced the ideas of relative velocity (known already to Aristoteles) and relative acceleration.

Robert Willis, a professor of natural end experimental philosophy, was teaching kinematics at Cambridge and his lectures were published in a book, *Principles of Mechanisms* (1841), which dealt with applied kinematics in the Ampere's sense and classified mechanisms on the basis on the change in directional relation and velocity ratio and identified different ways in achieving each function. Willis, who also was an accomplished historian of medieval architecture, further applied trigonometry and other mathematical methods for the kinematic analysis of specific mechanisms.

Among those who took up Willis ideas was Ferdinand Redtenbacher, professor at Karlsruhe Technical University, a great teacher who played an important role in the dissemination of Willis ideas. Moreover, he inspired one of his students, Franz Reuleaux, to become fascinated with kinematics.

Reuleaux was born to a family of master builders and his father was designer and builder of steam engines. He studied Engineering in Karlsruhe, natural sciences in Berlin and arts in Bonn. He taught a few years at the ETH in Zurich and moved to Berlin Technical University where he compiled a "mechanical alphabet" of mechanical devices of his-own. The advancing mechanization has created a great number of different machines and mechanisms and Reuleaux embarked in an attempt to systematize and classify them. His aim was to achieve the same uncompromisingly logical order in engineering as Linnaeus has done in biology. Known already from his 1854 book *Machine Elements* and the 1861 handbook *The Designer*, Reuleaux published in 1875 the *Theoretische Kinematik* which was translated immediately in English (1874-1876) by Kennedy under the title *Kinematics of Machinery*, where a complete treatment of kinematics of mechanism was presented. The book was supposed to be part of a trilogy, of which only one more book was published in 1900. Reuleaux, who was equally known to the cultural circles for his lectures and articles on "Culture and Technology", made extensive studies in the application of kinematics to the human body mechanics, translated poetry, wrote essays on art and books on a variety of subjects, had a great effect on the development of kinematics.

Rankine devoted a good part of his book *Machinery and Millworks* (1869) to machine kinematics. Rankine considered a machine to be made up of a *frame* and *moving pieces.* The frame is the fixed link in the language of Reuleaux. Rankine also made extensive use of the instantaneous centers of rotation. Although he introduced the *elementary combinations* of elements, his treatment was less general than that of Reuleaux. On the other hand he presented more specific results for particular forms of mechanisms.

Euler's ideas for planar motion found extensive application in Smith's (1825-1916) *Graphics* (1889) where extensive use was made of velocity and acceleration diagrams in mechanism analysis. In parallel, Burmester's (1840-1927) *Lehrbuch der Kinematic* (1888) adopted most of Reuleaux' methodology and also applied geometric methods for mechanism displacement, velocity and acceleration analysis. Aronhold in Germany and Kennedy in England extended the use of instantaneous centers of velocity while inventing the theorem of three centers. Gruebler (1851-1935) developed criteria for the movability of a linkage; this can be considered as an important step towards *number synthesis.* Finally, *dimensional synthesis* received attention in Burmester's work by geometric methods. It was further extended by Tschebytschev (1821-1894) and S. Roberts (1827-1913), who introduced triple generation of the coupler curves, and formalized much later by Freudenstein and Sandor.

Kinematicians were always fascinated by mechanisms for rectilinear motion. Watt considered his rectilinear motion mechanism his greatest achievement. Several mechanisms approximating linear motion were invented by Cartwright, White, Evans, Roberts and others. In 1864 Nicholas Peaucellier, a French military engineer, wrote a letter to the editor of a mathematical journal, proposing another rectilinear motion mechanism called "compas composé." There was immediate reaction from kinematicians including Pafnuti Tschebytschev (1821-1894) who had improved Watt's design. He tried to disprove mathematically Peaucellier's claim that his mechanism was producing an exact straight line. Proving that a kinematic chain, no matter how complex, could not produce an exact straight line became an obsession to him for the rest of his life,

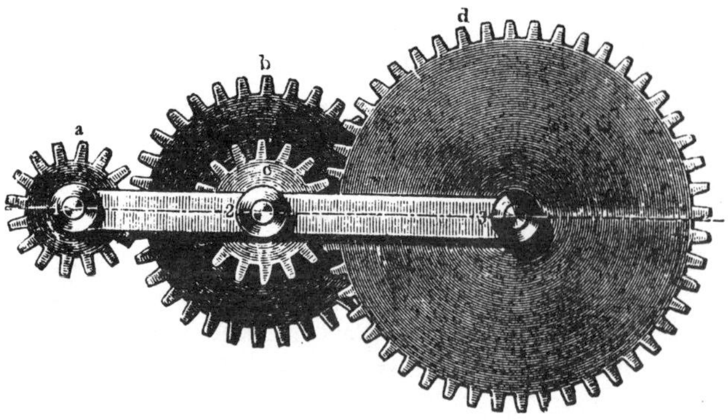

$$\ldots\ldots 2 \ldots\ldots\ldots b,a \ldots\ldots\ldots 1 \ldots\ldots\ldots 2$$

$$C^{+} \ldots | C_{z}^{+}, C_{z}^{+} \ldots | \ldots (C) \ldots || \ldots C_{=}^{-}$$

$$C^{+} \ldots | C_{z}^{+}, C_{z}^{+} \ldots | \ldots (C) \ldots || \ldots C_{=}^{-}$$

$$\ldots 2 \ldots\ldots\ldots c,d \ldots\ldots\ldots 3 \ldots\ldots\ldots 2$$

- - - - - - - - - - - - - - - -

$$C^{+} \ldots | \ldots \left\{ \begin{matrix} C_{z}^{+}, C_{z}^{+} \ldots | \ldots (C) \ldots \\ C_{z}^{+}, C_{z}^{+} \ldots | \ldots (C) \ldots \end{matrix} \right\} \ldots || \ldots C_{=}^{-}$$

$$\ldots\ldots\ldots\ldots\ldots c,d \ldots\ldots\ldots\ldots 3$$

Figure 14. Symboloic representation of a gear train [Reuleaux 1876)

but he was unable to find the proof.

The nineteenth century, in the words of Hartenberg and Denavit (1964),

... is the end of the formative period, of an era in which many principles had been uncovered, analytical methods established, and the road to synthesis opened. Subsequent growth of these ideas, no longer nurtured only in Europe, has been of such scope that no short overview is possible... The full story of kinematics of mechanisms, doing justice to the many who practiced the art of mechanisms and contributed to the science of kinematics, is yet to be written...

In the nineteenth century, as the Industrial Revolution in England flourished, the mechanization of production was widely adopted. This, did not always bring about changes in design methodology or production technology. The rising middle class demanded that their newly acquired wealth and social status be displayed in decorative, often ostentatious ways. While academics and intellectuals were debating aesthetic matters in industrial production, industry itself was striving for novelty by adopting the stylistic views of past cultures. These conflicting trends seriously impeded the natural and evolutionary development of form which, in many cases, was in sharp contrast to the development of product function.

A growing concern with the imbalance between function and form can be discerned in an editorial of Cole's Journal of Design, in 1850:

Design has a twofold relation, having in the first place a strict reference to utility in the thing designed; and secondarily, to beautifying or ornamenting that utility. The word design, however, with the many has become identified rather with its secondary than with its whole signification with ornament, as apart from and often as opposed to, utility. This was the reason for many of those great errors in taste which are observed in the works of many designers.

This view was further stressed by Colburn, a civil engineer, who in his 1871 book on locomotives, wrote *"none who aspired to become an engineer should encourage any play of imagination involving the forms or proportions of a mere mechanism..."*

SYMBOLIC REPRESENTATION-GROUP TECHNOLOGY

Most people believe that group technology is a contemporary issue but it really was developed some 200 years ago. Automation in the production of pulley blocks, introduced by Marc Brunel (1769-1849), a French-born engineer, generated a need for standardization of machine components. Christopher Pohlem (1661-1751), a Swedish engineer, introduced the *mechanical alphabet,* a collection of different components labeled with different letters of the alphabet. Every mechanism could be synthesized from elements of the mechanical alphabet just as a poem is comprised of simple letters. One of Pohlem's student-assistants was Carl Cronstedt (1709-1779) who described in a sketchbook about one-third of the elements of Pohlem's *mechanical alphabet.* He concentrated on the movements which can be generated, in particular by the *mighty five* simple elements, in contrast to Archimedes and the Alexandrians who dealt mainly with their force-amplifying features.

During this 17th and 18th centuries the idea of symbolic representation of a mechanism consisting of basic elements was used by clockmakers for the wheelwork of clocks, as for example, Outghred, 1677, Derham, 1696 and Alexandre, 1735. Babbage (1826) proposed the first complete symbolic description for the function and geometry of a linkage: an *Alphabet of Form* and an *Alphabet of Motion.* Willis (1841) attempted to improve Babbage's kinematic notation to make it applicable to mechanism design. It was Reuleaux however who presented in his *Kinematics of Machinery* book (1876) an extensive description and symbolic representation of mechanisms used in engineering. He defined the machine as *a combination of solid bodies, so arranged as to compel the mechanical forces of nature to perform work as a result of certain determinative movements.* Moreover, he dealt with very important aspects of kinematics, such as pairing and inversion of mechanisms. Reuleaux stated at the outset that he was concerned primarily with the theoretical aspects of machine kinematics and not applications. His aim was not to deal with the kinematic behavior of any particular machine but ... *to determine the conditions which are common to all machines ....* He did not consider it important to analyze, for example, the straight path generating linkages of Watt, Evans or Reichenbach, but rather to figure out how their inventors arrived at them with no knowledge of the underlying processes of thought. In this respect Reuleaux' work was the first to be devoted to modern kinematic synthesis and further to a symbolic representation of knowledge. Synthesis in the Reuleaux sense was limited to *type* synthesis. Systematic study of *type* and *dimensional* synthesis, although identified long before, were developed later. However, it is fair to consider Reulaeux as the father of *group technology.*

DESIGN METHODOLOGY

During the first half of the 19th century, following the Napoleonic wars, there was a systematic attempt to apply engineering science to the design analysis of machines. This movement matured about this time, culminating in Poncelet's *Introduction à la Mécanique Industrielle.* The focal point for the efforts to apply engineering science to mechanical design was at the Ecole Polytechnique, then as now a military school. The need to control the steam engine generated a wealth of investigations in kinematics and linkage designs (Hartenberg and Denavit 1964). The synthesis of linkages can be viewed as a formal design method. However, by its nature it cannot be considered as general mechanical design theory because it is limited to linkages.

In Germany, the von Humboldt model of universities entailed strong ties with industry, in contrast to the Ecole Polytechnique which was supported mainly by the French government. Consequently, in Germany there was a rapid dissemination of engineering science in mechanical engineering practice. J. Weisbach wrote a book on *Mechanics of Machinery and Engineering* which had a great impact on design analysis. F. Redtenbacher was another German professor of the same period who combined his teaching with a designer's job at Escher and Wyss Manufacturing Co. His books *Principien der Mechanik und des Maschinenbaues* (1852) and *Konstruktion fuer den Maschinenbau* (1862) included not only mechanics and strength of materials considerations in the design analysis of machines, but also introduced the concept of synthesis (konstruktion), a systematic methodology for finding design solutions and a set of general mechanical design principles, such as

a. sufficient strength,
b. sufficient resistance to wear,
c. sufficiently low friction,
d. optimum use of materials,
e. ease of manufacturing,
f. ease of maintenance,
g. simplicity.

Redtenbacher's principles are too contradictory and overlapping to be taken as a formal system of design axioms. A more abstract set of design principles was developed by his student Reuleaux (1854) who separately addressed considerations of function and form in the following two ground rules (Grundsaetze) of mechanical design:

a. *(Function) The design must provide a uniform satisfaction of the design requirements.* For example, in the case of loading by pressure, all parts should equally contribute to the support of the pressure.

b. *(Form) The form of the design embodiment must have the highest possible symmetry.*

The first principle can be regarded as an optimization principle. Equally loaded parts mean maximum utilization of the material, as Redtenbacher's principle (d). Stated in another form, when the design fails, it fails everywhere.

The second principle is essentially a minimum information principle. To an extent dependent on the particular design, it is not totally independent of the first principle, since symmetry results, in general, in lower manufacturing costs. It is not clear to what extent Reuleaux was driven to this principle by considerations of mere aesthetics or optimum design.

Reuleaux' design principles led to a balance between form and function considerations which had been separated since Alexandrian times and the decline of ancient Greek philosophy and science. Reuleaux saw another dimension in design:

> *Just as the poet contrasts the gentle and lovable Odyssean wanderers with the untamable Cyclops, the "lawless-thoughted monsters," so appear to us the unrestrained power of natural forces, acting and reacting in limitless freedom, bringing forth from the struggle of all against all their inevitable but unknown results, compared with the action of forces in the machine, carefully constrained and guided so as to produce the single desired result.*

CONCLUSION

It appears that certain fundamental ideas return with some regularity to the methodology of engineering design. It was pointed out above that group technology, symbolic manipulation and open ended design are not at all new ideas. The latter concept, promoted vigorously by the Accreditation Board of Engineering and Technology, was practiced extensively by the engineers of ancient times albeit, in a different way, since there is really no such thing as *open-ended design.* Engineering design, and design in general, is *single ended* by definition. It is often *open started,* in the sense that in the conceptual stage many design concepts can be considered. In fact, the latter is one of the distinguishing features of engineering from crafts. This is not a new discussion, the transformation from crafts to engineering design lasted for thousands of years. This was best described by Reuleaux (1874):

> *Formerly, the fundamental idea of alteration or extension was improvement, a word which says much in itself of the nature of the process. Now, on the other hand, we have a direct production of new things, a sudden bringing into being of heretofore completed machines. We see the beginnings of a perception that will someday apparently be*

universal among those who have to do with all classes of machinery. Upon this growing sense I believe that our polytechnic machine-instruction should act with increasing certainty.

ACKNOWLEDGMENTS

We gratefully acknowledge:

The extensive help on matters of style and substance by Professor Salvatore Sutera of Washington University, who read a draft of the manuscript and suggested numerous improvements.

Financial support through endowment funds made possible with a grant by the Danforth Foundation.

Computer programs written by Mr. Kevin Rintoul, Victoria, British Columbia, Canada, which facilitated systematic analysis of ancient texts.

REFERENCES

Aristoteles, Mhcanika. In *Aristotle, Minor Works*, Harvard University Press, 1936.

Babbage, C., *A Method for Expressing by Signs the Action of Machinery.* London, 1826.

Banu Musa bin Shakir brothers, *The Book of Ingenious Devices,* ca 850 A.D. Translated and annotated by D.R. Hill. Dodrecht-Holland: D. Reidel Publishing Co., 1979.

Benvenuto, E., An Introduction to the History of Structural Mechanics, Springer Verlag, New York, 1991,

Cohen, M.R., Drabkin, I.E., *A Source Book in Greek Science* Cambridge, Mass: Harvard Univ. Press, 1958.

Cotterell, B., Kamminga, J., Mechanics of Pre-industrial Technology, Cambridge University Press, Cambridge, 1990.

Dilke, O.A.W., *The Roman Land Surveyors, David and Charles*, Newton, Abbot, England, 1971.

Dimarogonas, A.D., *History of Technology.* Athens: Symmetry Pub. (in Greek), 1978.

Dimarogonas, A.D., The Origins of Vibration Theory, *Journal of Sound and Vibration*, 140(2), 181-189, 1990.

Dimarogonas, A.D., Mechanisms of the Ancient Greek Theater. *ASME Mechanisms Conference*, Phoenix, Arizona, 1992.

Dugas, R., *A History of Mechanics.* English translation. Neuchate, Switzerland: Editions du Griffon, and New York: Central Book Co., 1955.

Euler, L. *Mechanica sive motus scientia analytice exposita,* 1736-1742.

Euler, L., *Theoria motus corporum,* and 1775, Nov. Comm. Petrop., 1790.

Farrington, B., *Greek Science.* London: Hardsmondworth, 1944

Gladett, M., *Greek Science in Antiquity.* New York: Aberland-Schuman, Inc., 1955.

Herodotos, *Histories.* 5th Century *B.C.*

Hartenberg, R.S., Denavit, J., *Kinematic Synthesis of Linkages.* New York: Mc Graw-Hill, 1964.

Haselberger, L., The Construction plans for the Temple of Apollo at Didyma. *Scientific American*, December 1985, pp 126-132.

Heath, T., *A History of Greek Mathematics*, Oxford, The Clarendon Press, 1921. Dover, N.Y., 1981.

Hill, D., *A History of Engineering in Classical and Medieval Times*, Open Court Pub. Co., LaSalle, Illinois, 1984.

Hill, D.R., Mechanical Engineering in the Medieval Near East, Scientific American, pp 100-105, May 1991.

Ibn al-Razzaz al-Jazari, *The Book of Knowledge of Ingenious Devices*, 1206. Translated and annotated by D.R. Hill. Dodrecht-Holland: D. Reidel Publishing Co., 1974.

Johnson, W.A., Pliny the Elder and Standardized Roll Heights in the Manufacture of Papyrus, Classical Antiquity, January 1993.

Keller, W. *The Bible as History*, Bantam Books, New York, 1974, 1982.

Landels, J.G., *Engineering in the Ancient World*, University of California Press, Berkeley 1978.

Leupold, *Theatrum Machinarum,* 1724.

Mayr, O., *The Origins of Feedback Control,* English Translation, MIT Press, 1970.

Moll, C.L., Reuleaux, F., *Konstruktionslehre fuer den Maschinenbau.* Braunschweig: J. Vieweg, 1854.

Neugebauer, O., *Vorgriechische Mathematik,* Berlin, Springer Verlag, 1934.

Needham, J., Ling Wang, *Science and Civilization in China*, Cambridge, The Cambridge University Press, 1959.

Pappos of Alexandria, *Mathematical Collection*, 4th Century A.D.

Plinii, *Historia Naturalis, XIII*, Henry Bohn, London, 1865..

Pollux, Julius, *Onomasticon.*

Proclos Diadochos, 5th century A.D., *Commentary on Euclides Elements I*, Friedlein Ed.

Reuleaux, F., *Der Konstructeur.* Braunschweig: J. Vieweg, 1872.

Reuleaux, F., 1875. *Theoretische Kinematik*, translated in English by B.W. Kennedy. London 1876: Macmillan & Co.

Schopenhauer, A. *Die Welt als Wille und Forstellung*, P. Deussen Ed., Munich, 1911.

Schröedinger, E., *Nature and the Greeks*, Cambridge University Press, 1954.

Soedel, W., Foley, V., Ancient Catapults, *Scientific American*, 240, pp 120-8, 1979.

Strandh, S., *The History of the Machine*, Dorset Press, New York, 1989.

Theon of Smyrna, 2nd Century A.D., in Cohen and Drabkin, op. cit.

Vitruvius, Marcus, Polio, *On Architecture.* Cambridge, Mass: Harvard University Press, 1914.

Willis, R., *Principles of Mechanism,* 1841.

DE-Vol. 63, Vibrations of Mechanical Systems
and the History of Mechanical Design
ASME 1993

TRANSVERSE VIBRATIONS OF A ROLLER CHAIN DRIVE WITH A TENSIONER

Woosuk Choi and Glen E. Johnson
Department of Mechanical Engineering
and Applied Mechanics
Design Laboratory
University of Michigan
Ann Arbor, Michigan

ABSTRACT

A dynamic model for analysis of the performance of roller chain drive with a tensioner is presented. The model is based on the axially moving material model first given by Mote, but includes extensions that allow consideration of the effects of polygonal action, impact, and the periodic span length changes. Three equations of motion are obtained - one for the tensioner and one for each of the chain spans. The motion of the tensioner interacts with the motions of the chain spans through the chain tension, the contact angles, and the transverse displacement changes of the contact points. The effects of the periodic length change, tensioner stiffness, tensioner damping, tensioner position and external periodic load are investigated. Some of the results are compared to those obtained for a comparable chain drive without a tensioner. Solutions are obtained by a finite difference method.

NOMENCLATURE

c : chain span velocity
c_i : contact damping coefficient in impact
C_t : torsional damping coefficient of tensioner
I_2 : moment of inertia for driven sprocket system
I_t : effective moment of inertia of tensioner about its pivot point
k_i : contact stiffness coefficient in impact
K_t : torsional stiffness coefficient of tensioner
m_e : effective mass of tensioner
m_{eg} : effective mass of tensioner for gravity effect
m_i : effective mass in impact
m_l : mass per unit length of chain
M_{ts} : moment due to torsional spring
M_{td} : moment due to torsional damper
M_{ct} : moment due to chain tension
M_g : moment due to gravity
n_1, n_2 : driving and driven sprocket tooth numbers
r_1, r_2 : radius of driving and driven sprockets
r_t : radius of roller
P_c : chain tension
T_{ex} : magnitude of external periodic loading
T_p : tensioner preload torque
T_s : static tension
u, v : transverse displacements of chain span
V_{rel} : relative velocity of an engaging roller to sprocket surface
α_f: phase shift between engagement and disengagement
η : angle between relative velocity of roller and the sprocket surface
θ : tensioner arm angle
θ_0 : initial tensioner arm angle
ω_1 : angular velocity of driving sprocket
ω_2 : angular velocity of driven sprocket
ω_{ex} : frequency of external periodic loading

1. INTRODUCTION

Belt drives and chain drives are both commonly used methods of mechanical power transmission. Tensioners (or idlers) can be used on these two types of drives for several different reasons. In belt drives, tensioners are generally used to increase overall tension (to reduce slip). This has the additional benefit of increasing belt life. In roller chain drives, tensioners can improve drive performance by eliminating flop and sway on the slack side and reducing vibration and noise on the tight side (Marshek, 1986; Gould, 1990). In spite of the importance of the tensioner, there has not been much research reported in this area. Ulsoy, Whitesell and Hooven (1985) investigated a belt-tensioner system and reported Mathieu type instabilities due to the belt tension variations caused by a periodic external load. To date no theoretical investigations of chain drives with tensioners have been reported.

In this study the chain drive system comprises two chain spans and a torsional tensioner. The modeling of each chain span is based on an axially moving material (Mote, 1965). The modeling of polygonal action and impact is more completely described in reference (Choi and Johnson, 1993). Periodic length change for each chain span is also considered. The interaction between the tensioner and the chain spans and the performance of the chain drive with a tensioner are examined and the results are reported.

2. TRANSVERSE VIBRATION IN ROLLER CHAIN DRIVE WITH A TENSIONER

The chain drive system with a tensioner includes the driving and driven sprockets, tensioner assembly, and chain spans (see figure 1). The tensioner assembly consists of a tensioner arm, a tensioner sprocket, a torsional spring, and a torsional damper. In order to simplify the analysis, a circular disk is adopted as the tensioner sprocket and the contact line between the disk and the chain span is assumed to form a circular arc. The modeling of the tensioner and the chain span are done independently but they are coupled through the chain tension term, the inclination angles of the chain spans and the displacements of the contact points between the tensioner and the chain spans.

2.1 Dynamic Load

2.1.1 Speed Tensioning, Impact and External Periodic Load. The dynamic load due to speed tensioning is obtained in a way similar to the case of chain drive without a tensioner, i.e.,

$$D_s = \lambda m c^2 \tag{2-1}$$

The pulley support constant λ can be estimated as follows (Beikman et al., 1991).

$$\lambda = k_b/(k_b + k_s + k_g) \tag{2-2}$$

where $k_b = (EA/L)(\cos\psi_1 + \cos\psi_2)^2$, $k_s = k_t/l_t^2$, $k_g = T_s(\zeta_1 + \zeta_1)$, $\zeta_i = d(\cos\psi_i)/dx_s$, EA = chain strength, L = overall span length.

As for the impact, only the major impact between the engaging roller and the driving sprocket is considered. The dynamic load due to impact is determined in the same way as in the case of a chain drive without a tensioner (Choi and Johnson, 1993) and it is given by

$$D_{imp} = \frac{V_{rel}\sin 2\eta}{2\omega_d} e^{-\zeta\omega_n t} \{ (k_i - c_i \zeta \omega_n) \sin\omega_d t + c_i \omega_d \cos\omega_d t \} \quad (0 < t < \frac{\pi}{\omega_d}) \tag{2-3a}$$

$$D_{imp} = 0 \quad (\frac{\pi}{\omega_d} < t < \frac{\pi}{\omega_1 n_1}) \tag{2-3b}$$

where $V_{rel} = \omega_1 p_c$, p_c = chain pitch, $\eta = 35 + 240/n1 + \varepsilon$ deg., $\varepsilon << 1$, $\omega_n = (k_i/m_i)^{0.5}$, $\omega_d = \omega_n (1-\zeta^2)^{0.5}$, $\zeta = c_i/(2 (m_i k_i)^{0.5})$.

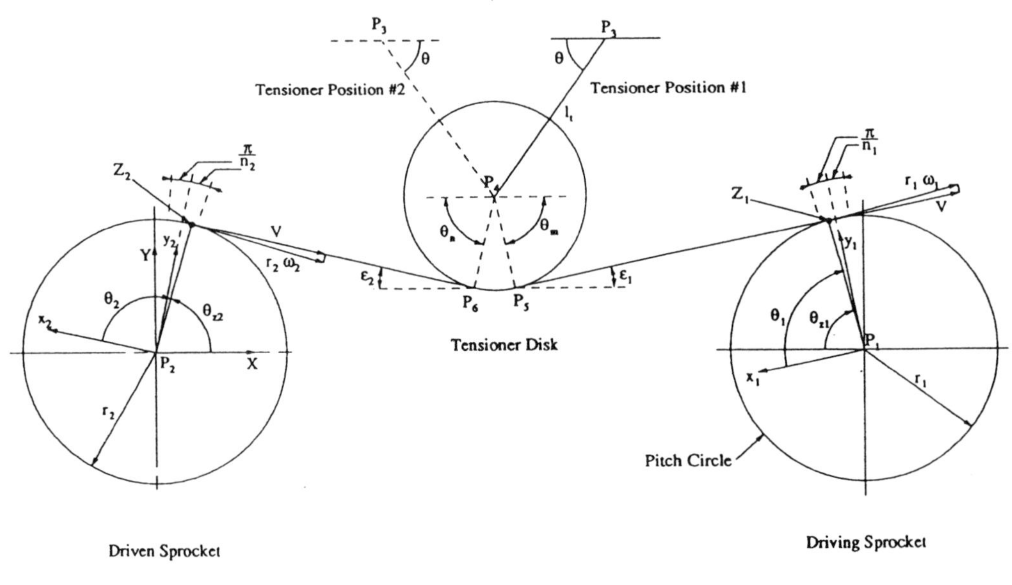

Figure 1 Configuration of Chain Drive With a Tensioner

External dynamic loads are included by a single trigonometric term.

$$D_{ext} = T_{ex} \cos \omega_{ex} t \qquad (2\text{-}4)$$

2.1.2 Polygonal Action. The major parameter related to polygonal action is the fractional pitch and it is determined by the positions of the tensioner and sprockets. The lengths of span #1 and #2 and the length of the arc formed around the tensioner disk are shown in figure 1. The fractional pitch is calculated based on the geometric information and there are two different general tensioner positions. One is the case where the tensioner pivot point stays on the right side of the vertical line passing through the center of the tensioner disk (tensioner position #1) and the other is the case where the pivot point resides on the left side of the vertical line (tensioner position #2).

$$f_p = \frac{L_1 + L_2 + (r_t + r_r)(\varepsilon_1 + \varepsilon_2) - i^*\text{pitch}}{\text{pitch}} \qquad (2\text{-}5)$$

where i = the integer that leads to $0 \le f_p \le 1$,

$$\varepsilon_1 = \frac{\pi}{2} + \tan^{-1}\left(\frac{P_{1Y} - P_{4Y}}{P_{1X} - P_{4X}}\right) - \tan^{-1}\left(\frac{L_1}{r_1 + r_t + r_r}\right),$$

$$\varepsilon_2 = \frac{\pi}{2} + \tan^{-1}\left(\frac{P_{2Y} - P_{4Y}}{P_{4X} - P_{2X}}\right) - \tan^{-1}\left(\frac{L_2}{r_2 + r_t + r_r}\right),$$

$$L_1 = \sqrt{S_1^2 - (r_1 + r_t + r_r)^2},$$

$$S_1 = \sqrt{(P_{1X} - P_{4X})^2 + (P_{1Y} - P_{4Y})^2},$$

$$L_2 = \sqrt{S_2^2 - (r_2 + r_t + r_r)^2},$$

$$S_2 = \sqrt{(P_{4X} - P_{2X})^2 + (P_{4Y} - P_{2Y})^2},$$

$$P_{4X} = P_{3X} - l_t \cos\theta \text{ (for tensioner position \#1)}$$

$$\text{or } P_{3X} + l_t \cos\theta \text{ (for tensioner position \#2)},$$

$$P_{4Y} = P_{3Y} - l_t \sin\theta .$$

The fractional pitch changes as time changes because ε_1 and ε_2 change. The fractional pitch was assumed to be constant in the case of a roller chain drive without a tensioner (Choi and Johnson, 1993). The phase angle (α_f) between engagement and disengagement is computed by multiplying one tooth angle of the driven sprocket by the fractional pitch.

The angular velocity relationship between the driving and driven sprockets is formulated by equating the chain velocities from the driving sprocket side and the driven sprocket side (figure 1). Using the x_1-y_1 coordinates and x_2-y_2 coordinates in the figure, the same equation used in the case of the chain drive without a tensioner is obtained. The way to obtain the dynamic load due to the polygonal action for the rigid four bar model (applicable at low speeds) is the same as presented in (Choi and Johnson, 1993) except that the constant fractional pitch is replaced with the time dependent fractional pitch given by equation (2-5). The angular motion of the driven sprocket and the dynamic load is given by

$$\theta_2 = \theta_{2int} \qquad \left(\theta_{2int} \le \frac{\pi}{2} + \frac{\pi}{n_2}\right) \qquad (2\text{-}6a)$$

$$\theta_2 = \theta_{2int} - \frac{2\pi}{n_2} \qquad \left(\theta_{2int} > \frac{\pi}{2} + \frac{\pi}{n_2}\right) \qquad (2\text{-}6b)$$

where $\theta_{2int} = \cos^{-1}\left\{\frac{r_1}{r_2}(\cos\theta_1 - \cos\theta_{10}) + \cos\theta_{20}\right\}$,

$$\theta_{10} = \theta_1(0) = \frac{\pi}{2} - \frac{\pi}{n_1}, \quad \theta_{20} = \theta_2(0) = \frac{\pi}{2} - \frac{\pi}{n_2} + \alpha_f,$$

$$\alpha_f = \frac{2\pi}{n_2} f_p.$$

$$D_{pol} = \frac{I_2}{r_2}\alpha_2 = \frac{I_2}{r_2^2 \sin\theta_2}(r_1 \omega_1^2 \cos\theta_1 - r_2 \omega_2^2 \cos\theta_2) \qquad (2\text{-}7)$$

where $\theta_1 = \omega_1 t + \frac{\pi}{2} - \frac{\pi}{n_1}$.

The way to obtain the dynamic load due to the polygonal action in case of the elastic four bar model (applicable at moderate and high speeds) is also the same as presented in (Choi and Johnson, 1993) except that the time-dependent fractional pitch is used instead of the constant fractional pitch. The angular movement of the driven sprocket and the dynamic load are given by

$$\theta_2 = \int_0^t \dot{\theta}_2 \, dt = \frac{\omega_1}{N^3 a_0^2}(q_0 t + q_1 \sin pt + q_2 \sin 2pt + q_3 \sin 3pt$$
$$+ q_4 \sin 4pt - p_0 t - p_1 \cos pt - p_2 \cos 2pt - p_3 \cos 3pt$$
$$- p_4 \cos 4pt + c_{int}) \qquad (2\text{-}8)$$

where $c_{int} = \frac{N^3 a_0^2}{\omega_1}\left(\alpha_f + \frac{\pi}{2} - \frac{\pi}{n_2}\right) + p_1 + p_2 + p_3$

$$+ p_4 - q_0 \tau \,\mathrm{dint}\left(\frac{t}{\tau}\right), \quad N = \frac{n_2}{n_1} = \frac{\frac{n_2 r_2}{\pi}\sin\left(\frac{\pi}{n_2}\right)}{\frac{n_1 r_1}{\pi}\sin\left(\frac{\pi}{n_1}\right)},$$

$p = n_1 \omega_1$, $a_0 = \frac{n_1 r_1}{\pi}\sin\frac{\pi}{n_1}$, dint(x) rounds the value of x to the nearest integer towards zero.

$$D_{pol} = k(x_2 - x_1 - L) = \frac{k\omega_1}{N p}\{(N a_1 - l_1)\sin pt +$$
$$m_1 \cos pt + 0.5(N a_2 - l_2)\sin 2pt +$$
$$0.5 m_2 \cos 2pt - m_1 - 0.5 m_2\} \qquad (2\text{-}9)$$

where $a_1 = \frac{2 a_0}{1 - n_1^2}$, $a_2 = \frac{2 a_0}{1 - 4 n_1^2}$.

The coefficients l_i, m_i, p_i, q_i, are described in Appendix I.

2.2 Transverse Displacements of End Points and Periodic Length Change

There are four end points - two for span #1 and the other two for span #2. This set includes the contact point between span #1 and the driving sprocket (end point #1), the contact point between span #1 and the tensioner disk (end point #2), the contact point between span #2 and the tensioner disk (end point #3) and the contact point between span #2 and the driven sprocket (end point #4).

The vertical displacements of end point #1 and end point #4 are calculated as follows.

$$u_n = r_1 (1 - \sin\theta_1) \qquad (2\text{-}10)$$

$$v_0 = r_2 (1 - \sin\theta_2) \qquad (2\text{-}11)$$

The angle of end point #4 is given by equation (2-6) at low speeds and by equation (2-8) at medium and high speeds. The vertical displacements of end points #2 and #3 are determined by the motions of the contact points between the tensioner disk and the chain spans. The computation of the contact points is based upon two angles (θ_m and θ_n in figure 1) formed by the chain spans and sprocket geometries.

$$P_{5X} = P_{4X} + (r_t + r_r)\cos\theta_m, \quad P_{5Y} = P_{4Y} - (r_t + r_r)\sin\theta_m \qquad (2\text{-}12)$$

$$P_{6X} = P_{4X} - (r_t + r_r)\cos\theta_n, \quad P_{6Y} = P_{4Y} - (r_t + r_r)\sin\theta_n \qquad (2\text{-}13)$$

where $\theta_m = \frac{\pi}{2} - \varepsilon_1$, $\theta_n = \frac{\pi}{2} - \varepsilon_2$.

Suppose P_5 has moved from $[P_5]_i$ to $[P_5]_{i+1}$ and P_6 has moved from $[P_6]_i$ to $[P_6]_{i+1}$ during one time step.

$$\alpha_a = [P_{5X}]_{i+1} - [P_{5X}]_i, \quad \beta_a = [P_{5Y}]_{i+1} - [P_{5Y}]_i \qquad (2\text{-}14)$$

$$\alpha_b = [P_{6X}]_{i+1} - [P_{6X}]_i, \quad \beta_b = [P_{6Y}]_{i+1} - [P_{6Y}]_i \qquad (2\text{-}15)$$

There are four different cases for each form of tensioner position (tensioner position #1 or tensioner position #2) in the computation of vertical displacements in terms of α_a, α_b, β_a, and β_b. The related angles and the changes of vertical displacements are shown in figure 2 for tensioner position #1 and the expressions for the changes (d_5 and d_6) are described in Appendix II.

$$[u_0]_{i+1} = [u_0]_i + [d_5]_i \qquad (2\text{-}16)$$

$$[v_n]_{i+1} = [v_n]_i + [d_6]_i \qquad (2\text{-}17)$$

where i, i + 1 represents time $i*\Delta t$ and $(i+1)*\Delta t$ respectively.

The length changes occur for both span #1 and span #2 since all the end points are moving. In the case of the chain drive without tensioner, the span changes its length abruptly whenever there is engagement or disengagement between chain links and the sprockets. When the tensioner is present, each span changes its length, not only abruptly at the engagement and the disengagement of chain links, but also gradually at other times. The span lengths in a chain drive with a tensioner are determined by the end points.

$$\text{spl1 (length of span \#1)} = \sqrt{(z_{1x} - P_{5x})^2 + (z_{1y} - P_{5y})^2} \qquad (2\text{-}18)$$

$$\text{spl2 (length of span \#2)} = \sqrt{(z_{2x} - P_{6x})^2 + (z_{2y} - P_{6y})^2} \qquad (2\text{-}19)$$

where $z_{1x} = P_{1x} - r_1\cos\theta_{z1}$, $z_{1y} = P_{1y} + r_1\sin\theta_{z1}$,

$z_{2x} = P_{2x} + r_2\cos\theta_{z2}$, $z_{2y} = P_{2y} + r_2\sin\theta_{z2}$,

$\theta_{z1} \approx \theta_1 - \varepsilon_1$, $\theta_{z2} \approx \pi - \varepsilon_2 - \theta_2$.

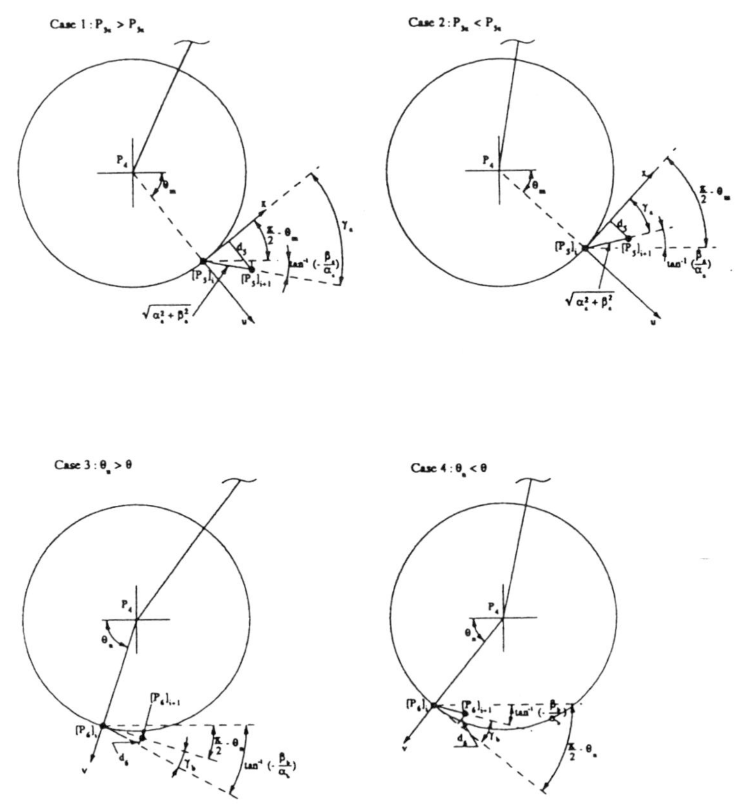

Figure 2 Angles Related with Contact Points Between Chain Span and Tensioner

2.3 Equations of Motion

The tensioner assembly is modeled as a single degree of freedom system with torsional damper and torsional spring and the tensioner disk is assumed to be a lumped mass at the end of the tensioner arm. The tensioner rotates about its pivot and the equation of motion is formulated by considering the moment equilibrium about the pivot point.

$$I_t \ddot{\theta} = M_{ts} + M_{td} + M_{ct} + M_g \quad (2\text{-}20)$$

There are four moments related to the pivot point of the tensioner and they are due to the torsional spring, torsional damping, chain span tension (see figure 3) and gravity.

$$M_{ts} = K_t (\theta_0 - \theta) + T_p \quad (2\text{-}21a)$$

$$M_{td} = - C_t \dot{\theta} \quad (2\text{-}21b)$$

M_{bt} =

$$P_c l_t\{\sin(\theta - \varepsilon_1) - \sin(\theta + \varepsilon_2)\} \text{ for tensioner position \#1} \quad (2\text{-}21c)$$

$$P_c l_t\{\sin(\theta - \varepsilon_2) - \sin(\theta + \varepsilon_1)\} \text{ for tensioner position \#2} \quad (2\text{-}21d)$$

where P_c(chain tension) $= T_s + D_s + D_{pol} + D_{imp} + D_{ext}$.

$$M_g = m_{eg}\, g\, l_t \cos\theta \quad (2\text{-}21e)$$

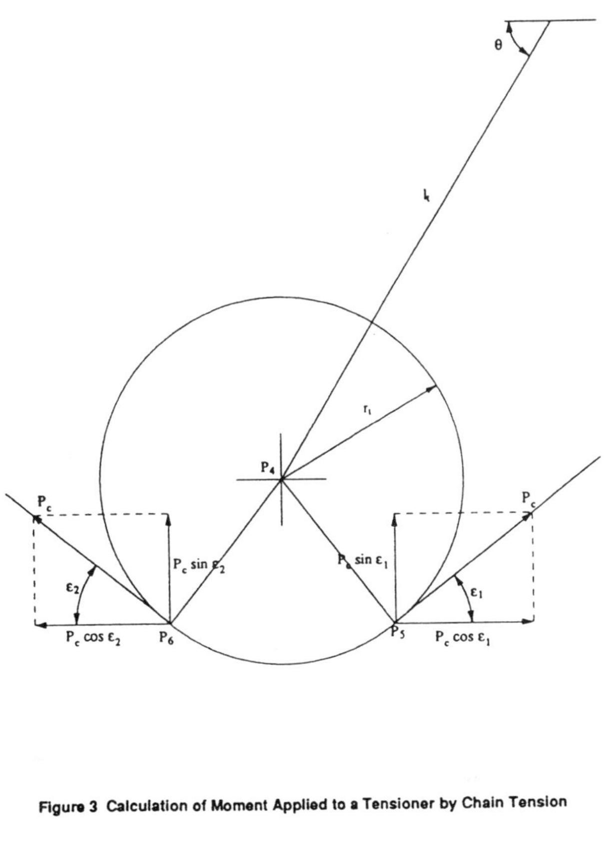

Figure 3 Calculation of Moment Applied to a Tensioner by Chain Tension

Substitution of equations (2-21) into (2-20) leads to the final form of the equation of motion.

$$m_e l_t^2 \ddot{\theta} + C_t \dot{\theta} + K_t \theta = K_t \theta_0 + P_t + P_c l_t\{\sin(\theta - \varepsilon_1) - \sin(\theta + \varepsilon_2)\} + m_{eg}\, g\, l_t \cos\theta \text{ for tensioner position \#1} \quad (2\text{-}22a)$$

$$m_e l_t^2 \ddot{\theta} + C_t \dot{\theta} + K_t \theta = K_t \theta_0 + P_t + P_c l_t\{\sin(\theta - \varepsilon_2) - \sin(\theta + \varepsilon_1)\} + m_{eg}\, g\, l_t \cos\theta \text{ for tensioner position \#2} \quad (2\text{-}22b)$$

The motion of the tensioner was represented by one ordinary differential equation which is coupled with the equations of motion for the chain spans through the tension term and inclination angles. Forward finite differencing was done about time to solve the equation.

There are two chain spans in a chain drive with a tensioner. The chain span in contact with driving sprocket is denoted as span#1 and the other is span#2. Because of the presence of the tensioner, sag can be neglected in both chain spans. The equation of motion for a chain drive without a tensioner can be used for each chain span. The transverse vibration of chain span #1 is represented by u and the transverse vibration of chain span #2 is represented by v.

$$\text{Span \#1}: m_l u_{,tt} + 2 m_l c\, u_{,tx} + m_l c^2 u_{,xx} + c_d(u_{,t} + c\, u_{,x}) - (P + P_t)\, u_{,xx} = 0 \quad (2\text{-}23)$$

$$\text{Span \#2}: m_l v_{,tt} + 2 m_l c\, v_{,tx} + m_l c^2 v_{,xx} + c_d(v_{,t} + c\, v_{,x}) - (P + P_t)\, v_{,xx} = 0 \quad (2\text{-}24)$$

where $P = T_s + D_c$, $P_t = D_{pol} + D_{imp} + D_{ext}$.

The boundary conditions for span #1 are given by the equations (2-10) and (2-16) and the boundary conditions for span #2 are given by the equations (2-11) and (2-17). The finite difference formulation followed the finite differencing scheme used in (Choi and Johnson, 1993).

3. COMPUTER SIMULATION

3.1 Time Step Control and Simulation Strategy

A nonuniform time step scheme was utilized to execute computation efficiently. The basic idea is to use smaller time steps near the instant of the impact between engaging roller and sprocket tooth and larger time steps away from that instant so that the effect of the impact is included while the total computation time is minimized without causing numerical instabilities. To achieve this objective, an exponential function was utilized.

The main objective of the simulation is to observe the change of the motions of the roller chain drive and the tensioner under different circumstances. In order to obtain the steady state response at an operating speed, the transient region during some amount of time from the beginning was ignored and the response after that was used to calculate the vibration amplitude. Next, the maximum amplitude from the equilibrium configuration of a chain span was chosen as a variable representing the vibration amplitude effectively over a wide range of operating speed. In order to represent the motion of the tensioner, two more variables were introduced - the average angle of the tensioner arm in vibration and the angular vibration amplitude of the arm about the average angle.

3.2 Simulation Results and Discussion

The chain drive system used in this simulation consists of number 40 chain, two 24 tooth sprockets and a tensioner. The effect of the moment of inertia for a driven sprocket system is investigated for a chain drive with a tensioner and the results are shown in figure 4. First of all, the multiple peaks that were observed in (Choi and Johnson, 1993) are no longer present. Instead, there is one peak corresponding to the average span length. This is because the span takes on the short span or long span length only instantaneously when a tensioner is present. The span length varies continuously between these two values during the remainder of the tooth period. The resonance of the tensioner is around 248 rpm and the angular vibration magnitude increases as the moment of inertia increases (especially around the resonances). This is because the increase in the moment of inertia induces an increase in the tension variation due to the polygonal action. The average angle of the tensioner decreases as the operating speed increases. This decrease is physically reasonable since the tension of the span increases as the speed increases due to speed tensioning and this tension increase pushes up the tensioner. The vibration amplitudes of span #1 and span #2 show similar trends. The vibration amplitude is not strongly affected by changes in the moment of inertia before the fourth resonance occurs, but the effect of changes in the moment of inertia is clear around the fourth resonance. In the case of the largest moment of inertia of figure 4 (solid line), even instability can even occur due to the parametric excitation.

The effects of tensioner stiffness are represented in figure 5. The location of the tensioner resonance increases as the stiffness increases and the amplitude at the resonance also increases as the stiffness increases since the excitation due to polygonal action increases as the speed increases. If the magnitude of the excitation

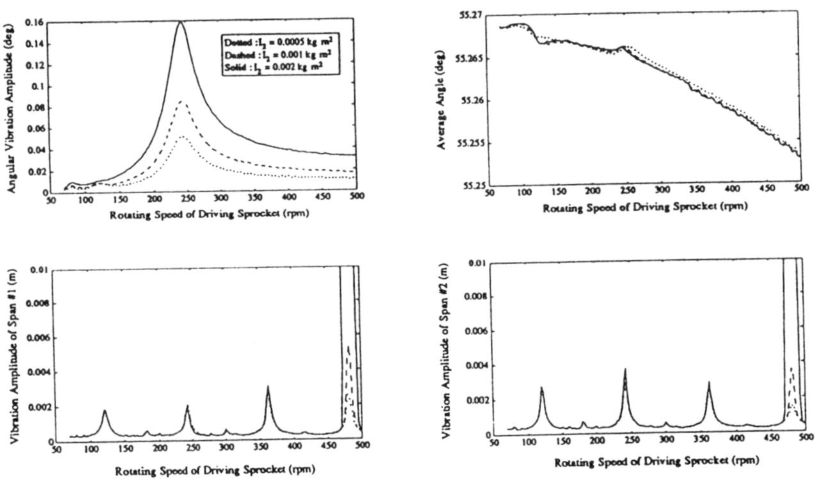

Figure 4 Effect of Moment of Inertia for Driven Sprocket System on Tensioner and Chain Span Vibrations (Chain Dirve : $n_1 = n_2 = 24$, cd = 0.4 m)

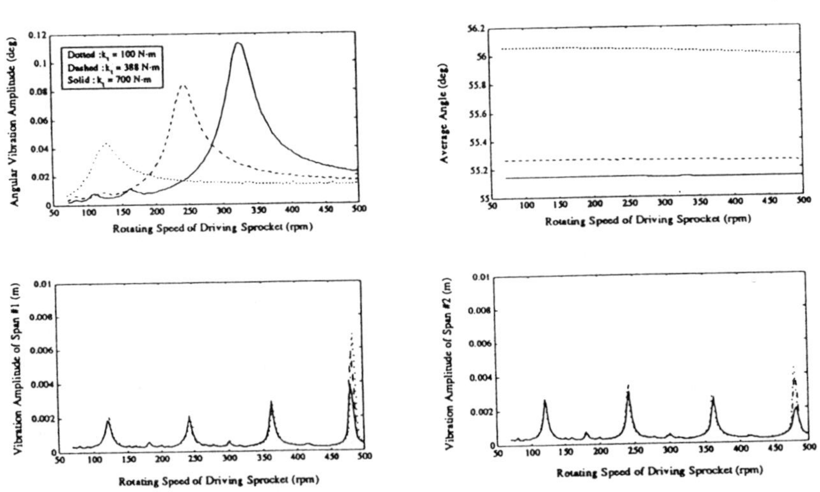

Figure 5 Effect of Tensioner Stiffness on Tensioner and Chain Span Vibrations (Chain Dirve : $n_1 = n_2 = 24$, cd = 0.4 m)

remains the same at any speed, the amplitude will decrease as the stiffness increases. The amplitude of span #1 around the fourth resonance increases as the stiffness decreases while below the resonance the effect of the tensioner stiffness is insignificant. The amplitude of span #2 also shows the same trend. The effect of tensioner damping is shown in figure 6. Three cases (underdamped, critically damped, and overdamped) were considered. The effect of the damping is noticeable around the fourth resonance for span #1 and around the second and the fourth resonances for span #2.

The position of the tensioner pivot point was changed in the vertical direction and the effects are shown in figure 7. One of the most interesting subjects for research in the area of roller chain drives is optimal tensioner position. Vertical changes in the location of the tensioner pivot point make a difference in the fractional pitch of the total chain span. Three cases were investigated. One is the system with 0.5258 initial fractional pitch (dotted line), another is the system with 0 initial fractional pitch (dashed line) and the other is the system with 0.6721 initial fractional pitch (dot-dashed) at 500 rpm. The initial fractional pitch is calculated at the equilibrium configuration for a given operating speed. As far as these three cases are concerned, the angular vibration amplitude is the largest for the system with 0.5258 fractional pitch and the smallest for the system with 0 fractional pitch. The effect on the vibration amplitude becomes large as the operating speed increases and there are some differences among the resonance positions because each span length changes as the vertical position of the tensioner pivot point changes.

The effect of the external periodic load on the case where the fractional pitch is 0 is also shown in figure 7. The effect is similar to that observed in the case of a roller chain drive without tensioner (Choi and Johnson, 1993) and the addition of the external periodic load with the frequency $\omega_{ex} = \omega_1$ reduces the vibration amplitude around the resonances. The angular vibration amplitudes increase

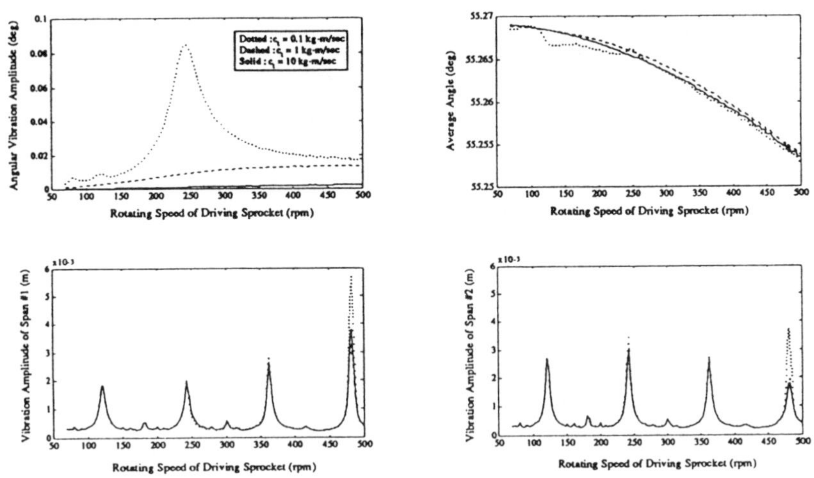

Figure 6 Effect of Tensioner Damping on Tensioner and Chain Span Vibrations (Chain Dirve : $n_1 = n_2 = 24$, cd = 0.4 m)

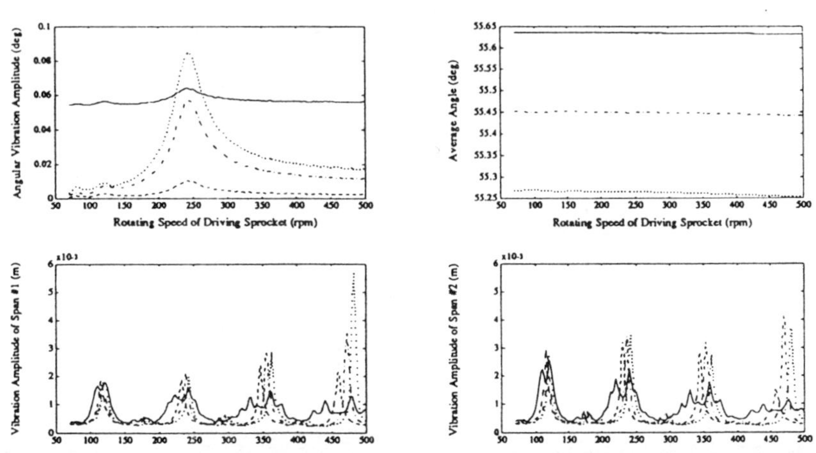

Figure 7 Effect of Tensioner Vertical Position Change and External Load on Tensioner and Chain Span Vibrations (Chain Dirve : $n_1 = n_2 = 24$, cd = 0.4 m)

Dotted : P_{yy} = 0.1 (f_p = 0.5258 at ω_1 = 500 rpm)
Dot-Dashed : P_{yy} = 0.115 (f_p = 0.6721 at ω_1 = 500 rpm)
Dashed : P_{yy} = 0.1303 (f_p = 0 at ω_1 = 500 rpm)
Solid : P_{yy} = 0.1303 w/ ext. load

due to the external periodic load.

Figure 8 shows the effect of the two different general tensioner position forms, corresponding to tensioner position #1 (P_3 = (0.25, 0.1)) and tensioner position #2 (P_3 = (.1702, 0.1)). The vibration amplitude with tensioner position #2 is larger than the amplitude with tensioner position #1 around the second resonance for span #1 and around the fourth resonance for span #1. It is interesting to notice that tensioner position #1 is preferred to tensioner position #2 in recent automotive serpentine belt designs.

The motions of the chain drive at medium operating speeds were also computed. Figure 9 shows the behavior of the chain drive when instabilities due to polygonal action in the elastic chain span occur. The system includes a relatively high inverse ratio of the tooth numbers (n_2/n_1 = 50/15) and small moment of inertia for the driven sprocket system (I_2 = .0005 kg m^2). As shown in the figure 9, the range of the vibration amplitude affected by the second instability is wider than the range affected by the first instability since there is a transverse resonance just above the second instability and the effect of the resonance is combined with the second instability.

As the last simulation, a comparison between a chain drive with and without a tensioner was done for a system with the same center distance (30 pitches) between the driving and driven sprockets (figure 10). The tensioner position was set to let the initial fractional pitch at the operating speed of 500 rpm be zero. For the first and second cases (T_s = 25 N and T_s = 50 N) there are instabilities due to the combined effect of parametric (polygonal action) and external (curvature and transverse displacement changes at both ends) excitations in a chain drive without tensioner, while there exist only relatively small vibrations in the chain drive with a tensioner. For the third and fourth cases (T_s = 100 N and T_s = 200 N), there are larger amplitudes at resonances but fewer number of peaks in the chain drive with a tensioner than in the chain drive without a tensioner since each chain span in the chain drive with a tensioner is

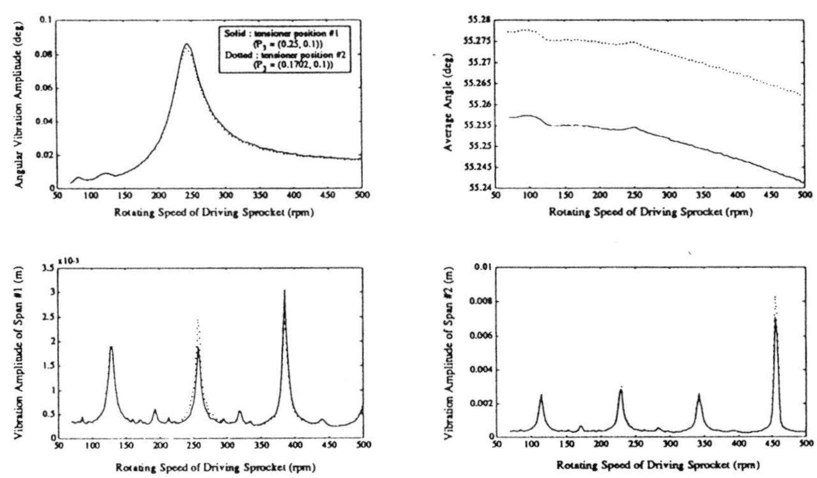

Figure 8 Effect of Two Different Tensioner Positions on Tensioner and Chain Span Vibrations (Chain Dirve : $n_1 = n_2 = 24$, cd = 0.4 m)

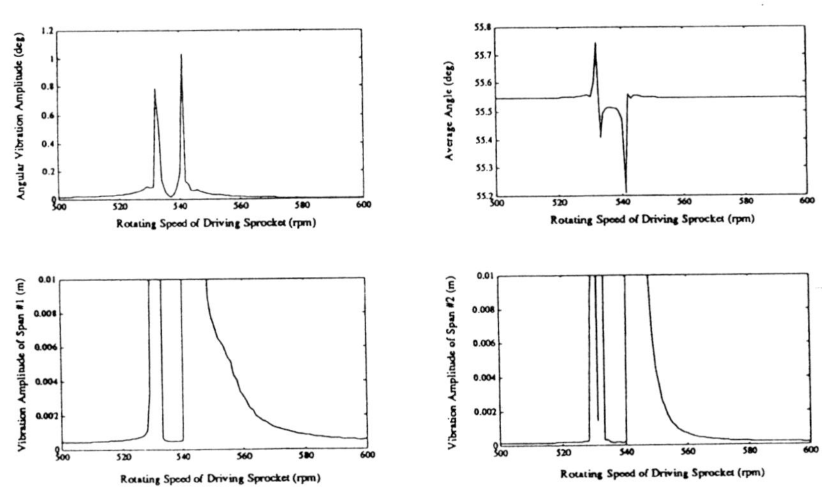

Figure 9 Behavior of Chain Drive with a Tensioner at Medium Operating Speeds (Chain Drive : $n_1 = 15$, $n_2 = 50$, I_2 = 0.0005 kg m^2)

smaller than the chain span in the chain drive without a tensioner.

4. SUMMARY AND CONCLUSION

A model for the chain drive with a tensioner was developed including polygonal action, impact, and periodic length change of each chain span. The tensioner interacts with the chain spans in several ways including chain tension. The periodic length change for each chain span does not produce multiple peaks around each order resonance as was the case for systems without tensioners. Instead, there is one peak at each resonance, corresponding to the average span length. This is because the tensioner causes the span length to vary continuously between the shortest and longest span lengths during one tooth period, rather than to abruptly take on either one value or the other. At low operating speeds, higher tensioner stiffness could reduce the vibration amplitude especially around high order resonances. But the location of the tensioner resonance increases and the vibration of the tensioner at the resonance also increases because the tension variation increases as the operating speed increases. Changes in the vertical location of the tensioner pivot point can be used to change the fractional pitch of the total chain span. The vibration amplitude can be reduced and sometimes instability can also be prevented by adjusting the vertical position to let the time dependent fractional pitch be approximately zero. At medium operating speeds, there is a danger of instabilities due to polygonal action. The possibility is generally low, but it increases as the inverse ratio of the tooth numbers increases and as the moment of inertia for the driven sprocket decreases. Compared to chain drives without tensioners, the tensioner can not only prevent instabilities for low static tension operations but can also reduce the number of peaks.

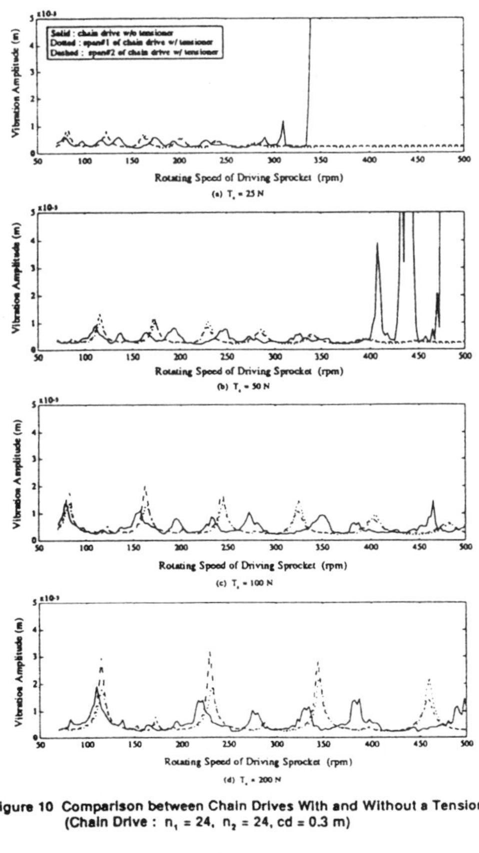

Figure 10 Comparison between Chain Drives With and Without a Tensioner (Chain Drive : n_1 = 24, n_2 = 24, cd = 0.3 m)

5. ACKNOWLEDGEMENTS

The authors gratefully acknowledge the support of the National Science Foundation through Grants No. MSM-88-12957 and MSS-8996293. We also wish to thank N. C. Perkins for his thoughtful suggestions during the early stage of the model development.

BIBLIOGRAPHY

Beikman, R. S., Perkins, N. C. and Ulsoy, A. G., 1991, "Equilibrium Analysis of Automotive Serpentine Belt Drive Systems Under Steady Operating Conditions", Proceedings of 22nd Midwestern Mechanics Conference, Oct., pp.533-534.

Choi, W. and Johnson, G. E., 1993, "Vibration of Roller Chain Drives at Low, Medium and High Operating Speeds", ASME 14th Biennial Conference on Mechanical Vibration and Noise.

Gould, Les, 1990, "Use an Idler to Solve Your Drive Problems", *Modern Materials Handling*, no. 45, Jun., p.75.

Marshek, K. M., 1986, "Standard Handbook of Machine Design - Chapter 32 Chain Drive", McGraw-Hill, pp.32.1 - 32.39.

Mote, C. D. Jr., 1965, "Some Dynamic Characteristics of Band Saws," Forest Products Journal, Jan., pp 37 -41.

Ulsoy, A. G., Whitesell J. E., and Hooven M. D., 1985, "Design of Belt-Tensioner Systems for Dynamic Stability", ASME Journal of Vibration, Stress, and Reliability in Design, vol.107, Jul., pp.282-290.

APPENDIX I

COEFFICIENTS OF EQUATIONS (2-8) AND (2-9)

The values of l_1, m_1, l_2, and m_2 are obtained from the following matrix equation.

$$X = (A^T A)^{-1} A^T F$$

where X is a 4×1 matrix whose elements are unknown variables (l_1, m_1, l_2, m_2).

The elements of the matrices in the above equation are shown in Appendix II of (Choi and Johnson, 1993)

The values of p_i and q_i are shown in Appendix III of (Choi and Johnson, 1993).

APPENDIX II

COMPUTATION OF TRANSVERSE DISPLACEMENTS OF CONTACT POINTS BETWEEN TENSIONER DISK AND CHAIN SPANS

There are four different cases to be considered in the computation of vertical displacements for each tensioner position (tensioner position #1 or tensioner position #2). For tensioner position #1, the related angles and the changes of vertical displacements are shown in figure 2 and computed as follows

Case 1 ($P_{3x} > P_{5x}$) : $\gamma_a = \frac{\pi}{2} - \theta_m + \tan^{-1}(-\frac{\beta_a}{\alpha_a}) \rightarrow$

$d_5 = \sqrt{\alpha_a^2 + \beta_a^2}\sin\gamma_a\ (\alpha_a > 0)$ or $-\sqrt{\alpha_a^2 + \beta_a^2}\sin\gamma_a\ (\alpha_a < 0)$

Case 2 ($P_{3x} < P_{5x}$) : $\gamma_a = \frac{\pi}{2} - \theta_m - \tan^{-1}(\frac{\beta_a}{\alpha_a}) \rightarrow$

$d_5 = \sqrt{\alpha_a^2 + \beta_a^2}\sin\gamma_a\ (\alpha_a > 0)$ or $-\sqrt{\alpha_a^2 + \beta_a^2}\sin\gamma_a\ (\alpha_a < 0)$

Case 3 ($\theta_n > \theta$) : $\gamma_b = -\frac{\pi}{2} + \theta_n + \tan^{-1}(-\frac{\beta_b}{\alpha_b}) \rightarrow$

$d_6 = \sqrt{\alpha_b^2 + \beta_b^2}\sin\gamma_b\ (\alpha_b > 0)$ or $-\sqrt{\alpha_b^2 + \beta_b^2}\sin\gamma_b\ (\alpha_b < 0)$

Case 4 ($\theta_n < \theta$) : $\gamma_b = \frac{\pi}{2} - \theta_n - \tan^{-1}(-\frac{\beta_b}{\alpha_b}) \rightarrow$

$d_6 = -\sqrt{\alpha_b^2 + \beta_b^2}\sin\gamma_b\ (\alpha_b > 0)$ or $\sqrt{\alpha_b^2 + \beta_b^2}\sin\gamma_b\ (\alpha_b < 0)$

In case 1 and case 2 the same expressions are derived but in case 3 and case 4 different expressions are presented. Finally the next vertical positions of end points #2 and #3 are obtained by adding the displacement changes calculated above to the present positions.

For tensioner position #2, the procedure to determine d_5 and d_6 is exactly the same as for the tensioner position #1.

Case 1 ($P_{3x} > P_{6x}$) : $\gamma_b = \frac{\pi}{2} - \theta_n - \tan^{-1}(-\frac{\beta_b}{\alpha_b}) \rightarrow$

$d_6 = -\sqrt{\alpha_b^2 + \beta_b^2}\sin\gamma_b\ (\alpha_b > 0)$ or $\sqrt{\alpha_b^2 + \beta_b^2}\sin\gamma_b\ (\alpha_b < 0)$

Case 2 ($P_{3x} < P_{6x}$) : $\gamma_b = \frac{\pi}{2} - \theta_n + \tan^{-1}(\frac{\beta_b}{\alpha_b}) \rightarrow$

$d_6 = -\sqrt{\alpha_b^2 + \beta_b^2}\sin\gamma_b\ (\alpha_b > 0)$ or $\sqrt{\alpha_b^2 + \beta_b^2}\sin\gamma_b\ (\alpha_b < 0)$

Case 3 ($\theta_m > \theta$) : $\gamma_a = -\frac{\pi}{2} + \theta_m - \tan^{-1}(\frac{\beta_a}{\alpha_a}) \rightarrow$

$d_5 = -\sqrt{\alpha_a^2 + \beta_a^2}\sin\gamma_a\ (\alpha_a > 0)$ or $\sqrt{\alpha_a^2 + \beta_a^2}\sin\gamma_a\ (\alpha_a < 0)$

Case 4 ($\theta_m < \theta$) : $\gamma_a = \frac{\pi}{2} - \theta_m - \tan^{-1}(\frac{\beta_a}{\alpha_a}) \rightarrow$

$d_5 = \sqrt{\alpha_a^2 + \beta_a^2}\sin\gamma_a\ (\alpha_a > 0)$ or $-\sqrt{\alpha_a^2 + \beta_a^2}\sin\gamma_a\ (\alpha_a < 0)$

DE-Vol. 63, Vibrations of Mechanical Systems
and the History of Mechanical Design
ASME 1993

VIBRATION OF ROLLER CHAIN DRIVES AT LOW, MEDIUM AND HIGH OPERATING SPEEDS

Woosuk Choi and Glen E. Johnson
Department of Mechanical Engineering
and Applied Mechanics
Design Laboratory
University of Michigan
Ann Arbor, Michigan

ABSTRACT

A model based on axially moving material is developed to study transverse vibration in roller chain drives. A unique feature of the work presented in this study is that impact, polygonal action and external periodic load have been included through chain tension and boundary conditions and periodic length change is also considered. The impact between the engaging roller and sprocket surface is modeled as a single impact between two elastic bodies and the modeling of the polygonal action is based on a four bar mechanism (rigid four bar at low speeds, elastic four bar at moderate and high speeds). At low and medium operating speeds, the system equation of motion for the chain span is expressed as a mixed type partial differential equation with time-dependent coefficients and time-dependent boundary conditions. At high operating speeds, the system equations of motion are two partial differential equations for transverse and longitudinal vibrations respectively and they are nonlinearly coupled. The effects on transverse vibration of center distance, the moment of inertia of the driven sprocket system, static tension, and external periodic load are presented and discussed. Solutions are obtained by a finite difference method and Galerkin's method.

NOMENCLATURE

c : chain span velocity
c_i : contact damping coefficient in impact
I_2 : moment of inertia of driven sprocket
k_i : contact stiffness coefficient in impact
L : chain span length
L_0 : the shorter of the two possible span lengths
L_{ct} : length of the common tangent
m_i : effective mass in impact
m_l : mass per unit length of chain
n_1, n_2 : driving and driven sprocket tooth numbers
p_c: chain pitch
r_1, r_2 : radius of driving and driven sprockets
r_t : radius of roller
T_{ex} : magnitude of external periodic loading
T_s : static tension
V_{rel} : relative velocity of a roller to sprocket surface
x_0 : relative displacement of roller to sprocket tooth surface after contact
α_f: phase shift between engagement and disengagement
η : angle between relative velocity of roller and the sprocket surface
λ : pulley support constant
θ : tensioner arm angle
θ_0 : initial tensioner arm angle
ω_1 : angular velocity of driving sprocket
ω_2 : angular velocity of driven sprocket
ω_{ex} : frequency of external periodic loading

1. INTRODUCTION

Chain drives are characterized by the discrete nature of the chain links and sprocket teeth. Compared to belt drives, this discrete nature makes the chain drive unique and gives both advantages and disadvantages. The advantages include no-slip between the chain and the sprocket and lower loads on the shaft bearings. The disadvantages include noise and vibration (Jackson and Moreland, 1955). These undesirable characteristics have motivated several researchers to investigate the behavior of chain drives including their transverse vibration.

The transverse vibration of a chain strand can be excited by both internal and external stimuli. Periodic torsional loading (Wang, 1991) and imbalance in the drive are external sources; polygonal action and roller-tooth impacts are internal sources. Polygonal action occurs due to the fact that the chain, lying on the sprocket, forms a polygon rather a circle. This chordal nature leads to a fluctuating motion. Previous studies include the effects of center distance and sprocket-tooth numbers on the fluctuation of the velocity ratio of two sprockets (Morrison, 1952; Bouillon and Tordian, 1965; Chen and Freudenstein, 1988) and the derivation of the dynamic load due to polygonal action (Mahalingam, 1956; Turnbull and Fawcett, 1975). Another internal source corresponds to the impact between the engaging roller and the driving sprocket and it is due to the velocity of the roller relative to the sprocket surface as the roller seats. Research on the impact has been done on the effective mass (Ryabov, 1968; Turnbull and Fawcett, 1973; Chew, 1985), the contact stiffness (Fawcett and Nicol, 1979) , the elastic collision (Kim, 1990), and the intensity of impact interacting with transverse vibration (Wang et al., 1992). Also the impact force was measured experimentally in (Conwell et al., 1992a, 1992b, 1992c).

Studies of the transverse vibration of the chain have been based on both continuous and discrete models. One type of discrete model encompasses a string loaded with lumped masses (Binder, 1956; Tsujioka, 1976; Veikos and Freudenstein, 1992). An alternate discrete model is based on a series of links (Turnbull and Fawcett, 1973; Shimizu and Sueoka, 1975; Kim, 1990). The continuous model is typically based on a uniform string. Transverse vibration and stability analysis of a uniform string was addressed in (Mahalingam, 1957; Naguleswaran and Williams, 1968). Another continuous model that has been applied to study chain drives is the axially moving string. A recent study investigated the stability of an axially moving string subject to the dynamic load due to polygonal action (Ariaratnam and Asokanthan, 1987). The discrete model is best suited for the investigation of the detailed dynamics of a chain drive (like the contact mechanics), but it is less well suited for the study of the vibration of a general chain drive (including more than two sprockets) due to the complexity of modeling.

Thus a primary goal of this research is to extend the continuous model to incorporate the dynamic load due to polygonal action, the dynamic load due to impact, the periodic length change of chain span and the transverse displacement excitations at the end points of the chain span. The new model will be used to study the performance (including transverse vibration) of chain drives at low, medium, and high operating speeds.

2. TRANSVERSE VIBRATION IN ROLLER CHAIN DRIVE

2.1 Dynamic Load

2.1.1 Speed Tensioning, Impact, and External Periodic Load. The dynamic loads in roller chain drives are due to speed tensioning, polygonal action, impact and external loading. The dynamic load due to speed tensioning is expressed for a general axially moving material as follows.

$$D_c = \lambda\, m_l\, c^2 \tag{2-1}$$

The value of λ represents the relative stiffness of an axially moving material compared to that of the supporting structure. For a system with one chain span and two sprockets, λ can range from 0 to 1 (Mote, 1965). When λ is one, it means that the axially moving material is very stiff compared to the structure. In the case of roller chain drives, λ is closer to one than to zero for most cases because of the high stiffness of the chain span.

The dynamic load due to impact depends on the effective mass taking part in the impact, the relative impact velocity, roller-tooth contact stiffness and contact damping. Based on previous work (Ryabov, 1968; Turnbull and Fawcett, 1973; Chew, 1985), the effective mass is assumed to be the mass of two chain links in the present study. The stiffness coefficient is calculated by linearizing the nonlinear function developed by Dubowsky and Freudenstein (1971). The damping is assumed to be viscous and the coefficient is taken as 0.1 for normally lubricated chain. The relative velocity is calculated approximately as follows.

$$V_{rel} = 2r_1\omega_1\sin(\frac{\pi}{n_1}) = \omega_1\, p_c \tag{2-2}$$

The effective mass can be considered as a small mass m_i moving with velocity V_{rel} and about to contact the surface of a sprocket tooth with an angle η between the surface and the direction of V_{rel} as shown in figure 1. The angle is determined by the contact point of the engaging roller. The contact point of the engaging roller remains very close to the transition point staying in the working curve close to the border between the seating curve and the working curve (Kim and Johnson, 1992a, 1992b). Thus the angle is derived using the theoretical pressure angle representing the border.

$$\eta = 35 + \frac{240}{n_1} + \varepsilon \text{ (degrees)} \tag{2-3}$$

where $\varepsilon << 1$.

A single impact between two elastic bodies is assumed to occur while the roller is in contact with the sprocket tooth surface. The component of the contact force in the chain span direction is the dynamic load due to the impact. The detailed derivation is in Appendix I. Defining a tooth period as $2\pi/(\omega_1 n_1)$, the dynamic load for one tooth period becomes

$$D_{imp} = \frac{V_{rel}\sin 2\eta}{2\omega_d} e^{-\zeta\omega_n t}\{ (k_i - c_i\,\zeta\,\omega_n)\sin\omega_d t + c_i\,\omega_d\cos\omega_d t \}\quad (0 < t < \frac{\pi}{\omega_d}) \tag{2-4-a}$$

$$D_{imp} = 0 \quad (\frac{\pi}{\omega_d} < t < \frac{\pi}{\omega_1 n_1}) \tag{2-4-b}$$

The shape of D_{imp} is shown in figure 1.

Chains are often used in applications where the external loads are cyclic in nature, e.g. camshaft drives and intermittent motion machinery. The frequency of the load can vary from one cycle in a number of sprocket revolutions, to many cycles per revolution. In this paper the external load is modeled as previously suggested by Ulsoy, Whitesell, and Hooven (1985). The load comprises one cosine term.

$$D_{ext} = T_{ex} \cos \omega_{ex} t \qquad (2\text{-}5)$$

2.1.2 Polygonal Action. Two models were developed to compute dynamic load due to polygonal action. One is based on a rigid four bar mechanism and is used for the analysis of chain drives at low operating speeds. The other is based on an elastic four bar mechanism and is used for the analysis of chain drives at medium and high operating speeds.

a) Dynamic load based on rigid four bar model (applicable to low speeds). In order to calculate the dynamic load, the angular velocity of the driven sprocket is needed. It is assumed that the angular velocity of the driving sprocket is constant. The fractional pitch is necessary for the representation of the phase between engagement and disengagement. The fractional pitch is defined as

$$f_p = \frac{L_{ct} - i^* p_c}{p_c} \qquad (2\text{-}6)$$

where i is the integer which makes $0 \le f_p \le 1$.

The additional fractional pitch that is dependent on the operating speed (Kim, 1990) is not considered in this study. The angular velocity relationship between the driving and driven sprockets is given by the following equation (see figure 2) when chain velocity is assumed not to be affected by span vibration.

$$\omega_2 r_2 \sin \theta_2 = \omega_1 r_1 \sin \theta_1 \qquad (2\text{-}7)$$

Feasible domains :

$$\frac{\pi}{2} - \frac{\pi}{n_1} \le \theta_1 \le \frac{\pi}{2} + \frac{\pi}{n_1}, \quad \frac{\pi}{2} - \frac{\pi}{n_2} \le \theta_2 \le \frac{\pi}{2} + \frac{\pi}{n_2}$$

where $\theta_1 = \omega_1 t + \frac{\pi}{2} - \frac{\pi}{n_1}$, $\theta_2(0) = \frac{\pi}{2} - \frac{\pi}{n_2} + \alpha_f$, $\alpha_f = \frac{2\pi}{n_2} f_p$.

In order to solve equation (2-7) for ω_2, the value of θ_2 must be obtained first. Equation (2-7) is integrated to get

$$\theta_2 = \theta_{2int} \qquad (\theta_{2int} \le \frac{\pi}{2} + \frac{\pi}{n_2}) \qquad (2\text{-}8\text{-}a)$$

$$\theta_2 = \theta_{2int} - \frac{2\pi}{n_2} \qquad (\theta_{2int} > \frac{\pi}{2} + \frac{\pi}{n_2}) \qquad (2\text{-}8\text{-}b)$$

where $\theta_{2int} = \cos^{-1}\left\{\frac{r_1}{r_2}(\cos \theta_1 - \cos \theta_{10}) + \cos \theta_{20}\right\}$,

$\theta_{10} = \theta_1(0) = \frac{\pi}{2} - \frac{\pi}{n_1}$, $\theta_{20} = \theta_2(0) = \frac{\pi}{2} - \frac{\pi}{n_2} + \alpha_f$.

With the obtained value of θ_2, ω_2 can be computed using equation (2-7). The angular acceleration of the driven sprocket (α_2) is obtained by differentiating equation (2-7) and the dynamic load is calculated considering the moment equilibrium about the driven sprocket shaft.

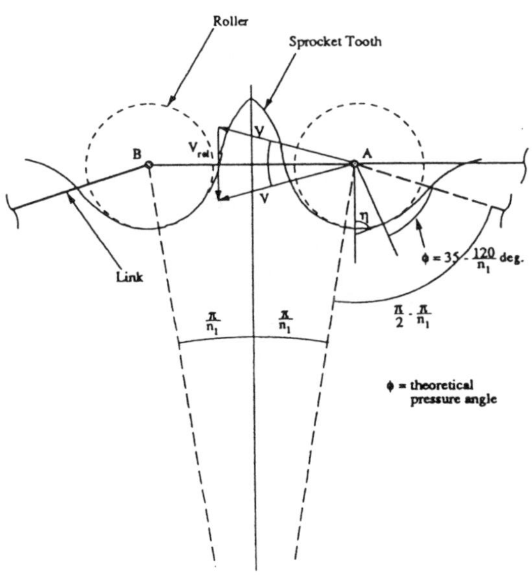

(a) Position of Rollers and Contact Angles at the Instant of Impact

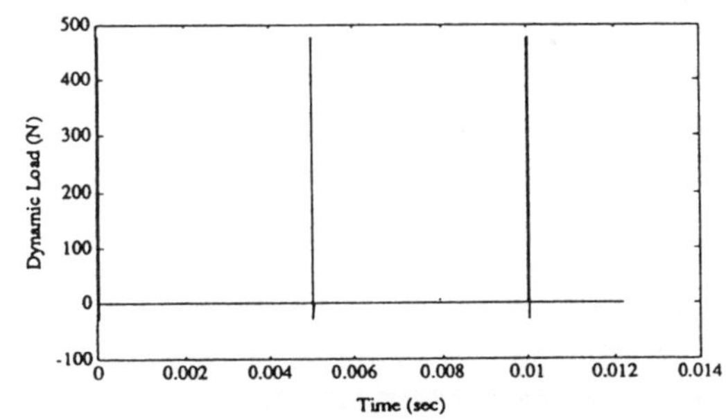

(b) Shape of Dynamic Load in Wide Time Range

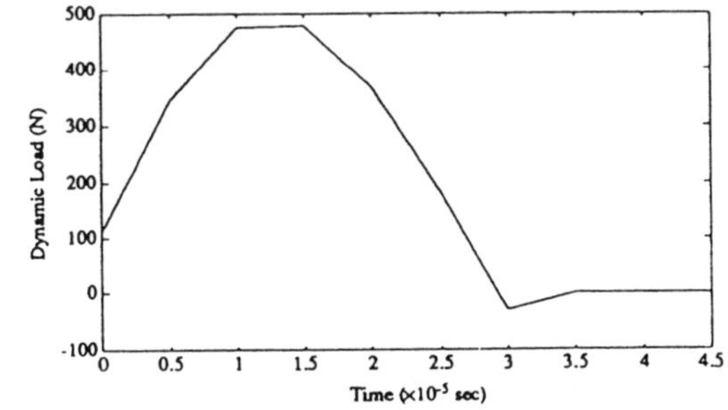

(c) Shape of Dynamic Load near Impact

Figure 1 Impact Phenomenon in Roller Chain Drive

$$\alpha_2 = \frac{1}{r_2 \sin \theta_2} (r_1 \omega_1^2 \cos \theta_1 - r_2 \omega_2^2 \cos \theta_2) \tag{2-9}$$

$$D_{pol} = \frac{I_2}{r_2} \alpha_2 \tag{2-10}$$

The dynamic load due to polygonal action is shown in figure 3 for three cases - (a) $n_1 = n_2$, (b) $n_1 > n_2$, and (c) $n_1 < n_2$. The magnitude of the dynamic load increases as the operating speed increases. The shape of the load is determined by the inverse ratio of the sprocket tooth numbers ($= n_2/n_1$).

<u>**b) Dynamic load based on elastic four bar model.**</u> As the operating speed increases, the elasticity of chain span becomes important. The following analysis of the dynamic load takes the elasticity into consideration. It is assumed that the driving sprocket runs at a constant speed of ω_1. The reference axes, taken at O_1, are parallel and perpendicular to the common tangent to the pitch circles of the sprockets. The change of slope of the chain is neglected. Since the quantity $r_1 \sin \theta_1$ is a function of tooth period τ ($= 2\pi/(n_1\omega_1)$), it can be expressed as a Fourier series.

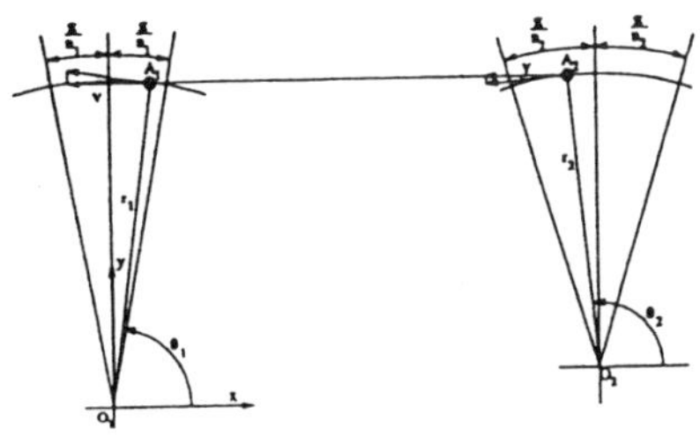

Figure 2 Equivalent Four-Bar Mechanism for The Calculation of Angular Velocity of Driven Sprocket

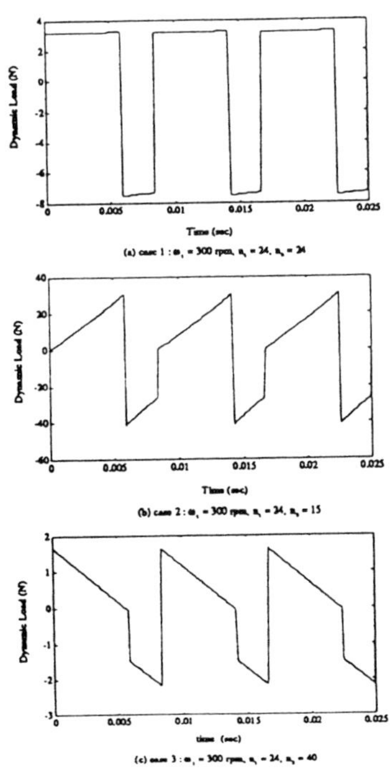

Figure 3 Dynamic Load Due to Polygonal Action for Rigid Chain Span

$$r_1 \sin \theta_1 = \frac{n_1 r_1}{\pi (1 - i^2 n_1^2)} \sin \frac{\pi}{n_1} + \sum_{i=1}^{\infty} \frac{2 n_1 r_1}{\pi (1 - i^2 n_1^2)} \sin \frac{\pi}{n_1} \cos ipt \tag{2-11}$$

where $p = n_1 \omega_1$.

It is approximated up to second harmonic terms considering that the values of the coefficients decrease rapidly as the order of the harmonic increases.

$$r_1 \sin \theta_1 \approx a_0 + a_1 \cos pt + a_2 \cos 2pt \tag{2-12}$$

where $a_0 = \frac{n_1 r_1}{\pi} \sin \frac{\pi}{n_1}$, $a_1 = \frac{2 a_0}{1 - n_1^2}$, $a_2 = \frac{2 a_0}{1 - 4 n_1^2}$.

The speed of A_1 is given by

$$\dot{x}_1 \approx - \omega_1 (a_0 + a_1 \cos pt + a_2 \cos 2pt)$$

and by integration

$$x_1(t) \approx - \omega_1 (a_0 t + \frac{a_1}{p} \sin pt + \frac{a_2}{2p} \sin 2pt) + x_1(0) \tag{2-13}$$

In a similar manner it can be shown that

$$r_2 \sin \theta_2 \approx N a_0 + b_1 \cos pt + c_1 \sin pt + b_2 \cos 2pt + c_2 \sin 2pt \tag{2-14}$$

where $b_1 = \frac{2 N a_0}{1 - n_2^2} \cos n_2\alpha$, $c_1 = - \frac{2 N a_0}{1 - n_2^2} \sin n_2\alpha$,

$b_2 = \frac{2 N a_0}{1 - 4 n_2^2} \cos 2n_2\alpha$, $c_2 = - \frac{2 N a_0}{1 - 4 n_2^2} \sin 2n_2\alpha$,

$$N = \frac{n_2}{n_1} = \frac{\frac{n_2 r_2}{\pi} \sin(\frac{\pi}{n_2})}{\frac{n_1 r_1}{\pi} \sin(\frac{\pi}{n_1})}.$$

Equation (2-14) was derived based on the assumption that the speed of the driven sprocket is constant. In fact the speed is subject to small fluctuations but it can be shown that the error involved in equation (2-14) is negligible.

Let the speed of A_2 be given by

$$\dot{x}_2 = - \frac{\omega_1}{N} (N a_0 + l_1 \cos pt + m_1 \sin pt + l_2 \cos 2pt + m_2 \sin 2pt) \tag{2-15}$$

Integrating equation (2-15) and substituting boundary conditions, we get

$$x_2(t) = -\frac{\omega_1}{N}\left(N a_0 t + \frac{l_1}{p} \sin pt - \frac{m_1}{p} \cos pt + \frac{l_2}{2p} \sin 2pt - \frac{m_2}{2p} \cos 2pt\right) - \frac{\omega_1}{N p}\left(m_1 + \frac{m_2}{2}\right) + x_1(0) + L \tag{2-16}$$

Now the angular velocity of the driven sprocket is calculated noting that b_1, b_2, c_1 and c_2 are small quantities and the angular acceleration is obtained by differentiating the angular velocity.

$$\dot{\theta}_2 = -\frac{\dot{x}_2}{r_2 \sin \theta_2} \approx \frac{\omega_1}{N^3 a_0^2}(N a_0 + l_1 \cos pt + m_1 \sin pt + l_2 \cos 2pt + m_2 \sin 2pt)(N a_0 - b_1 \cos pt - c_1 \sin pt - b_2 \cos 2pt - c_2 \sin 2pt) \tag{2-17}$$

$$\ddot{\theta}_2 = \frac{d}{dt}(\dot{\theta}_2) \tag{2-18}$$

For oscillations of the driven system about the axis of rotation, the equation of motion is

$$I_2 \ddot{\theta}_2 + k (x_1 + L - x_2) r_2 \sin \theta_2 = 0 \tag{2-19}$$

where I_2 : moment of inertia of driven system about axis of rotation, $k = \frac{AE}{L}$, AE = strength of chain span.

Substituting equations (2-13), (2-14), (2-16), and (2-18) into equation (2-19) and letting each term related with each harmonic be zero, a set of algebraic equations are obtained.

$$A X = F \tag{2-20}$$

where X is a 4×1 matrix whose elements are unknown variables (l_1, m_1, l_2, m_2).

The elements for the matrices are shown in Appendix II.

There are four unknowns and nine equations, resulting in an over-determined system. Since there is no exact solution, an approximate solution is achieved using a least squares method.

$$X = (A^T A)^{-1} A^T F \tag{2-21}$$

Based on the values of l_1, m_1, l_2, m_2 obtained above, the dynamic load is calculated as follows.

$$D_{pol} = k (x_2 - x_1 - L) = \frac{k \omega_1}{N p} \{(N a_1 - l_1) \sin pt + m_1 \cos pt + 0.5 (N a_2 - l_2) \sin 2pt + 0.5 m_2 \cos 2pt - m_1 - 0.5 m_2\} \tag{2-22}$$

θ_2 is obtained by integrating equation (2-17) symbolically.

$$\theta_2 = \int_0^t \dot{\theta}_2 \, dt = \frac{\omega_1}{N^3 a_0^2}(q_0 t + q_1 \sin pt + q_2 \sin 2pt + q_3 \sin 3pt + q_4 \sin 4pt - p_0 t - p_1 \cos pt - p_2 \cos 2pt - p_3 \cos 3pt - p_4 \cos 4pt + c_{int}) \tag{2-23}$$

where $c_{int} = \frac{N^3 a_0^2}{\omega_1}(\alpha_f + \frac{\pi}{2} - \frac{\pi}{n_2}) + p_1 + p_2 + p_3 + p_4 - q_0 \tau \, \mathrm{dint}(\frac{t}{\tau})$,

dint(x) rounds the value of x to the nearest integer towards zero.

The coefficients including q_i, and p_i are shown in Appendix III.

In addition, there is an abrupt length change in chain span except when the length of the common tangent is equal to an integral number of pitches. The length increases by one pitch when the chain link disengages from the driven sprocket and the length decreases by one pitch when the chain link engages with the driving sprocket. This length change causes a change in the longitudinal stiffness of the span thus introducing a sudden change in the dynamic load due to polygonal action.

$$k_1 = AE / L_0, \quad k_2 = AE / (L_0 + p_c) \tag{2-24a,b}$$

where L_0 is the shorter of the two possible span lengths.

Therefore there are two system characteristic resonances in the dynamic load corresponding to the two cases ($k = k_1$ and $k = k_2$) in

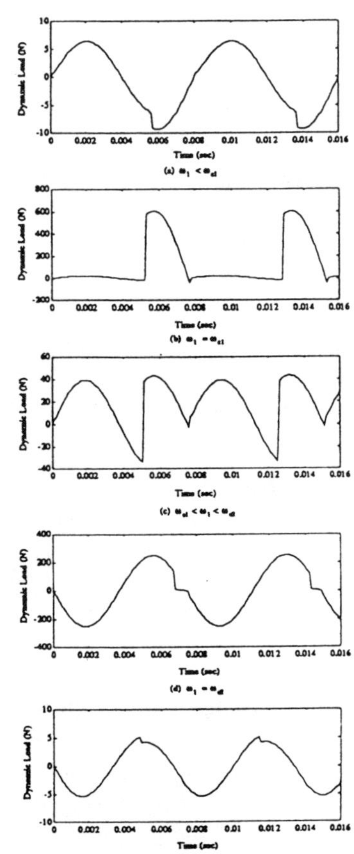

Figure 4 Dynamic Load Due to Polygonal Action for Elastic Chain Span

contrast with Mahalingam's results (1958) where there was only one. Furthermore, the magnitude of the dynamic load is limited around the resonances while the magnitude was unlimited in Mahalingam's results. The magnitude change of the dynamic load along with the change of the operating speed and various shapes of the dynamic loads are shown in figure 4. The two resonances (ω_{c1} and ω_{c2}) can cause Mathieu type dynamic instabilities of chain drive system. The existence and the positions of the peaks, if they exist, are determined by operating speed and system characteristics including longitudinal stiffness of chain span, the inverse ratio of the sprocket tooth numbers, and moment of inertia of the driven sprocket system.

2.2 Transverse Displacements of End Points and Periodic Length Change

There are two end points for a chain span, one associated with each sprocket. Each end point is moving along an arc formed by one tooth angle ($2\pi/n_1$ or $2\pi/n_2$) and there is generally a phase shift between the two periodic motions of the end points. The vertical motions (u_n, u_0) of the end points are approximated as

$$u_n = r_1 (1 - \sin \theta_1) \qquad (2\text{-}25)$$

where $\theta_1 = \omega_1 t + \frac{\pi}{2} - \frac{\pi}{n_1}$.

$$u_0 = r_2 (1 - \sin \theta_2) \qquad (2\text{-}26)$$

The value of θ_2 is determined by either equation (2-8) or (2-23). Both u_n and u_0 are shifted by $- 0.5\, r_2 (1 - \cos (\pi/n_2))$ in order to consider the proper shapes of the displacement excitations in the computer simulation. As mentioned in section 2.1.2.(b), there is a length change in general chain drive operation. The length change occurs either at the engagement of the chain link or the disengagement. When θ_1 goes beyond its limit $\pi/2 + \pi/n_1$ (engagement), the free span length decreases by one pitch. When θ_2 goes beyond its limit $\pi/2 + \pi/n_2$ (disengagement), the span length increases by one pitch. This length change affects not only the shape of the dynamic load due to polygonal action but also the transverse vibration of the chain drive.

2.3 Equations of Motion

It is proposed here that a chain drive system can be modeled as a moving uniform string and that the discrete nature (polygonal action and impact phenomenon) can be incorporated into the system equation of motion through the tension term and the boundary conditions.

2.3.1 General Equations of Motion. The general equations of motion for an axially moving chain span are derived in two-dimensional space using Hamilton's principle following the similar procedure in (Perkins, 1987). Strain energy, gravitational potential energy and kinetic energy are formulated. There is external work due to both the static tension and the dynamic tensions (speed tensioning, polygonal action, impact and external periodic load) and the work is done to the chain span through both ends.

Two equations of motion are derived for both longitudinal and transverse motions.

$$\begin{aligned} u_1 :\ & \{(P^e + E A^e \varepsilon)\, a_1\}_{,S} - (P^e + E A^e \varepsilon)\, \kappa^e a_2 - \rho A^e g\, l_1 \\ &= \{\rho A^e (u_{1,t} + c^e a_1)\}_{,t} + \{\rho A^e c^e (u_{1,t} + c^e a_1)\}_{,S} \\ &\quad - \rho A^e c^e \kappa^e (u_{2,t} + c^e a_2) \end{aligned} \qquad (2\text{-}27)$$

$$\begin{aligned} u_2 :\ & \{(P^e + E A^e \varepsilon)\, a_2\}_{,S} + (P^e + E A^e \varepsilon)\, \kappa^e a_1 - \rho A^e g\, l_2 \\ &= \{\rho A^e (u_{2,t} + c^e a_2)\}_{,t} + \{\rho A^e c^e (u_{2,t} + c^e a_2)\}_{,S} \\ &\quad + \rho A^e c^e \kappa^e (u_{1,t} + c^e a_1) \end{aligned} \qquad (2\text{-}28)$$

where S : arc length coordinate measured along the chain span,

u_1, u_2 : the displacements along and perpendicular to the chain span from the equilibrium respectively

$\varepsilon = u_{1,S} - \kappa^e u_2 + \frac{1}{2} \{ (u_{1,S} - \kappa^e u_2)^2 + (u_{2,S} + \kappa^e u_1)^2 \}$,

$a_1 = 1 + u_{1,S} - \kappa^e u_2$, $a_2 = \kappa^e u_1 + u_{2,S}$,

$P^e = \sqrt{\{(P_0 - m_l c^{e2})^2 + (m_l g \bar{S}^e)^2\}} + m_l c^{e2}$,

$\kappa^e = \dfrac{m_l g (P_0 - m_l c^{e2})}{(P_0 - m_l c^{e2})^2 + (m_l g \bar{S}^e)^2}$,

P^e : tension of chain span at equilibrium configuration,
P_0 : tension of the lowest point of chain span at equilibrium configuration,
κ^e : curvature at equilibrium configuration.
$\bar{S}^e$: span length from the lowest point at equilibrium

There are two geometric boundary conditions ($u_2(0,t)$, $u_2(L^e,t)$) calculated through the kinematic analysis of chain span and two natural boundary conditions related with the longitudinal motion.

$u_2(0,t)$, $u_2(L^e,t)$: prescribed
$-(P^e + E A^e \varepsilon)\, a_1 + P_1 = 0$ at $S^e = 0$
$-(P^e + E A^e \varepsilon)\, a_1 + P_2 = 0$ at $S^e = L^e$

2.3.2 Equation of Motion for Chain Drive at Low and Medium Speeds. In most applications of chain drives at low operating speeds, there exists a small sag[1] in the chain span. The nonlinear equations of motion about the equilibrium configuration are obtained from (2-27) and(2-28). The superscript e is dropped out from here forward.

$$\begin{aligned} u_1 :\ & \{(E A \varepsilon + (P + E A \varepsilon)(u_{1,S} - \kappa u_2)\}_{,S} - (P + E A \varepsilon)\, \kappa\, (u_{2,S} \\ &+ \kappa u_1) = [m_l \{u_{1,t} + c\, (u_{1,S} - \kappa u_2)\}]_{,t} + [m_l c\, \{u_{1,t} + \\ &c\, (u_{1,S} - \kappa u_2)\}]_{,S} - m_l c \kappa \{u_{2,t} + c\, (u_{2,S} + \kappa u_1)\} \end{aligned} \qquad (2\text{-}29)$$

$$\begin{aligned} u_2 :\ & \{(P + E A \varepsilon)\, (u_{2,S} + \kappa u_1)\}_{,S} + \kappa \{E A \varepsilon + (P + E A \varepsilon) \\ &(u_{1,S} - \kappa u_2)\} = [m_l \{u_{2,t} + c\, (u_{2,S} + \kappa u_1)\}]_{,t} + [m_l c\, \{u_{2,t} + \\ &c\, (u_{2,S} + \kappa u_1)\}]_{,S} + m_l c \kappa \{u_{1,t} + c\, (u_{1,S} - \kappa u_2)\} \end{aligned} \qquad (2\text{-}30)$$

[1] The boundary between small and large sags is determined in terms of sag-span ratio (r = sag/ span length). As a rule of thumb, small sag theory holds when r is less than 1/8.

With the small sag assumption, the curvature and the tension at equilibrium are approximated up to order of δ ($\sim m_l\, g\, L_c/(P_0 - m_l c^2)$).

$$P = \sqrt{\{(P_0 - m_l c^2)^2 + (m_l\, g\, \overline{S})^2\}} + m_l c^2 \approx P_0 \tag{2-31}$$

$$\kappa = \frac{m_l\, g\, (P_0 - m_l c^2)}{(P_0 - m_l c^2)^2 + (m_l\, g\, \overline{S})^2} \approx \frac{m_l\, g}{P_0 - m_l c^2} = \frac{\delta}{L_c} \tag{2-32}$$

where L_c = total curved span length.

Assume that the chain stretches in a quasi-static manner (E A >> P) and linearize the equation of motion for the u_1 component to get

$$(E\, A\, \varepsilon)_{,S} = 0 \tag{2-33}$$

Thus the dynamic component of chain tension becomes

$$f(t) = E\, A\, \varepsilon = E\, A\, (u_{1,S} - \kappa\, u_2) \tag{2-34}$$

At both ends,

$$f(t) = P_t = E\, A\, (u_{1,S} - \kappa\, u_2) \tag{2-35}$$

The tangential displacement is obtained by integrating equation (2-35) and removing the rigid body mode.

$$u_1 = \kappa\, (\int_0^S u(\eta,t)\, d\eta - \int_0^{L_c} u(\eta,t)\, d\eta) + \frac{P_t}{E\, A}(S - L_c) \tag{2-36}$$

Linearizing the equations of motion for u_2 component about the quasi-static solution and including viscous damping in transverse direction gives

$$m_l\, u_{2,tt} + 2\, m_l\, c\, u_{2,tS} + m_l\, c^2\, u_{2,SS} + c_d(u_{2,t} + c\, u_{2,S}) - \{(P + P_t)\, u_{2,S}\}_{,S} - \kappa\, P_t = 0 \tag{2-37}$$

where $P = T_s + \lambda\, m\, c^2$, $P_t = D_{pol} + D_{imp} + D_{ext}$.

There are both parametric and external excitations in the system equation of motion.

In the case of negligible sag, the equation of motion becomes

$$m_l\, u_{2,tt} + 2\, m_l\, c\, u_{2,tx} + m_l\, c^2\, u_{2,xx} + c_d(u_{2,t} + c\, u_{2,x}) - \{(P + P_t)\, u_{2,x}\}_{,x} = 0 \tag{2-38}$$

Equations (2-37) and (2-38) are not simple parabolic or hyperbolic partial differential equations but mixed type partial differential equations with time dependent coefficients and time dependent boundary conditions. In order to solve the equations of motion, a finite difference method was adopted. Central finite differencing was done about time and mixed finite differencing was done about the spatial coordinate. The mixed finite differencing combines central differencing at low operating speeds and backward differencing at high speeds (Ulsoy et al., 1985). Accordingly, the first derivative for the spatial coordinate is expressed as

$$\frac{\partial u}{\partial S} = \frac{1 - s^*}{2\, \Delta S}\, u_{i+1,j} + \frac{s^*}{\Delta S}\, u_{i,j} - \frac{1 + s^*}{2\, \Delta S}\, u_{i-1,j} \tag{2-39}$$

where i = subscript for spatial coordinate,
j = subscript for time, $s^* = \dfrac{c}{\sqrt{T_s/m_l}}$.

2.3.3 Equations of Motion for Chain Drive at High Operating Speeds

As the excitation frequency (e.g. tooth frequency) related to the operating speed approaches the first resonance of longitudinal motion, the quasi-static assumption no longer holds and equations (2-29) and (2-30) should be solved simultaneously. When a chain drive is operated at such high speeds, sag is usually not allowed in practice. There are two cases where the simultaneous solution of the two equations is required. The first case occurs when the tooth frequency dependent loads have a significant effect on the coupling between the transverse and longitudinal equations of motion and the tooth frequency is near the first longitudinal resonant frequency. The second case occurs when the frequency of the external load is an integral multiple of the frequency of rotation of the driving sprocket, this frequency is near the first longitudinal resonance, and the amplitude of the external excitation is large enough to make the coupling term significant. This case does not happen unless the operating speed of the chain drive is very high. In this study sprocket dynamics are not considered and there is no direct dynamic coupling between the chain span and the sprocket.

The nonlinear equations of motion of a chain with an arbitrary sag are expressed about the equilibrium configuration by equations (2-29) and (2-30). Several modifications are required to obtain the equations of motion for the chain without sag, specifically

$$\kappa = 0,\ S = x,\ a_1 = 1 + u_{1,x},\ a_2 = u_{2,x},$$
$$\varepsilon \approx u_{1,x} + \frac{1}{2}\, u_{2,x}^2 \quad (E\, A >> P)$$

With those modifications and the addition of viscous damping in the transverse direction, the equations of motion and boundary conditions can be obtained. The mode shapes for an axially stationary string are taken as the comparison functions for the transverse vibration. Suitable comparison functions for the longitudinal vibration are not readily determined, so admissible functions defined by the longitudinal vibration modes of a free-free rod are used to discretize the equations of motion.

$$u_1 = \sum_{j=1}^{N} \psi_j(t) \cos\frac{j\, \pi\, x}{L} \tag{2-40}$$

$$u_2 = \sum_{j=1}^{N} \phi_j(t) \sin\frac{j\, \pi\, x}{L} + \frac{L - x}{L}\, g(t) + \frac{x}{L}\, h(t) \tag{2-41}$$

where $g(t) = u_2(0,t)$, $h(t) = u_2(L,t)$.

Since the admissible functions are used in the discretization process, the discretization from the variational form is necessary.

The variational form for the chain drive without sag is obtained from the original Hamilton's formulation with several modifications including viscous damping in the transverse direction.

$$
\begin{aligned}
u_1 : \int_{t_1}^{t_2} \Big[\int_0^L & \{-m_l (u_{1,tt} + 2\,c\,u_{1,xt} + c^2\,u_{1,xx}) + E\,A\,u_{1,xx} \\
& + E\,A\,u_{2,x}\,u_{2,xx}\}\,\delta u_1\,dx + \{E\,A\,(u_{1,x} + \tfrac{1}{2}\,u_{2,x}^2) - P_t\}\,\delta u_1\,|_{x=0} \\
& + \{-E\,A\,(u_{1,x} + \tfrac{1}{2}\,u_{2,x}^2) + P_t\}\,\delta u_1\,|_{x=L}\Big]\,dt = 0 \qquad (2\text{-}42)
\end{aligned}
$$

$$
\begin{aligned}
u_2 : \int_{t_1}^{t_2} \int_0^L & [-m_l (u_{2,tt} + 2\,c\,u_{2,xt} + c^2\,u_{2,xx}) + P\,u_{2,xx} \\
& - c_d\,(u_{2,t} + c\,u_{2,x}) + E\,A\,\{u_{1,xx}\,u_{2,x} + u_{1,x}\,u_{2,xx} \\
& + 1.5\,u_{2,x}^2\,u_{2,xx}\}]\,\delta u_2\,dx = 0 \qquad (2\text{-}43)
\end{aligned}
$$

The final form of the discretized equations of motion was obtained by substituting expressions (2-40) and (2-41) into equations (2-42) and (2-43). The discretized equations of motion are a set of nonlinearly coupled second order ordinary differential equations and the Runge-Kutta fourth order method was used as an integration scheme.

3. COMPUTER SIMULATION

3.1 Time Step Control and Simulation Strategy

A nonuniform time step scheme was developed for efficient computation. The basic objective of the scheme is to use smaller time steps near the instant of the impact (between engaging roller and sprocket tooth) and larger time steps away from that instant so that the effect of the impact is included while the total computation time is minimized without causing numerical instabilities. To achieve this objective, an exponential function was utilized.

The main objective of the simulation is to observe the change of the vibration amplitude of the roller chain drive under different circumstances. In order to obtain the steady state response at each operating speed, the transient region during some amount of time from the beginning was ignored and the response after that was used to calculate the vibration amplitude. The chain motion is complex. For purpose of the discussion, it is desirable to characterize this complex motion by a single descriptor. Our experience indicates that a good choice for this descriptor is the maximum amplitude from the chain span equilibrium position. This variable can illustrate the vibration effectively over a wide range of operating speeds.

3.2 Simulation Results and Discussion

A specific chain drive system was selected for detailed study. It comprises one number 40 chain and two 24 tooth sprockets. The center distance ranged from 19.7 pitches to 20.5 pitches. First simulation was done to investigate the effect of periodic length change and the center distance on the vibration response. Figure 5 shows both effects. The solid line represents the simulation results obtained considering a center distance as an average span length and the dashed line represents the results obtained including periodic length change in chain span. The difference becomes clearer as the operating speed approaches high order resonances. When there is no consideration of the length change, one peak of the vibration amplitude occurs at each order resonance for the center distance. In contrast, when the periodic length change is considered, multiple peaks can occur around each order resonance except in the case where the center distance is an integral number of the pitches. The reason for the multiple peaks is the effect of the periodic length change combined with the dynamic load due to the polygonal action. The effect of center distance becomes clear when the system with 20 pitch center distance (system A) is compared with the system with 20.5 center distance (system B). Around the first resonance the response of the system A was clearly amplified but the one of the system B was not. This result is in agreement with the results of previous discrete approaches (Turnbull and Fawcett, 1972; Kim, 1990). The reason is that the transverse displacements at both ends of system A are in phase, thus exciting the first mode effectively. However, the displacements of system B are out of phase and can not efficiently excite the first mode. As shown in the observation of vibration amplitudes at the first resonance, center distance has a major influence on the positions of higher resonances and the vibration amplitudes. Since the amplitude at a specific operating speed is affected significantly by the center distance, the optimal

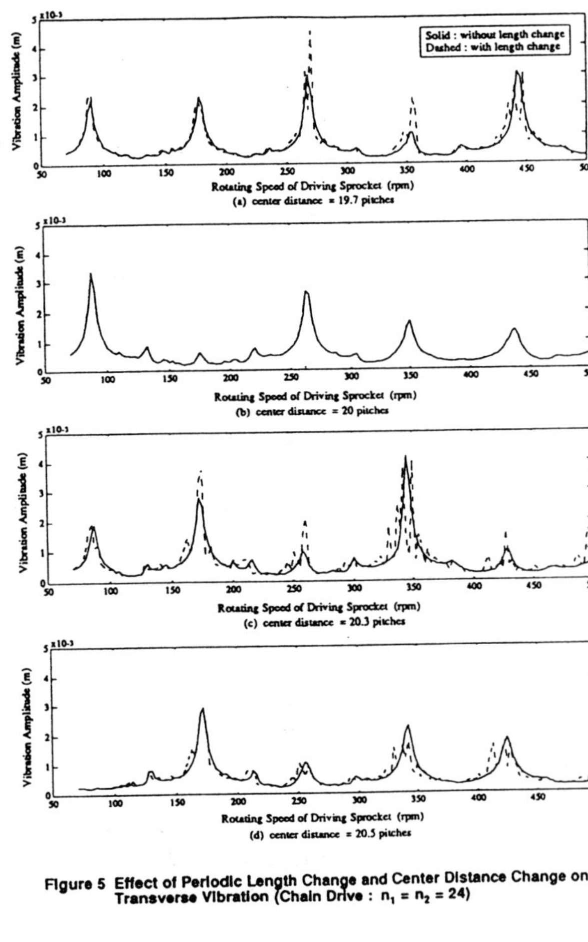

Figure 5 Effect of Periodic Length Change and Center Distance Change on Transverse Vibration (Chain Drive : $n_1 = n_2 = 24$)

center distance should be decided by the speed at which the system is operated.

In addition to the transverse displacement excitations at both ends of the chain span, there are other intrinsic excitations which correspond to dynamic loads due to polygonal action and impact. The effect of polygonal action is dominant at low and medium speeds while the effect of impact is dominant at high operating speed. The magnitude of the tension variation due to the polygonal action depends on the moment of inertia of the driven sprocket system and operating speed. The first effect of the parametric excitation due to the tension variation appears as the increase of the amplitudes around the resonances as shown in figure 6a and also there is a danger of instabilities just as shown in figure 6b. Figure 7 shows the effect of static tension. As the magnitude of the static tension decreases, the possibility of the instabilities due to the dynamic load by the polygonal action increases. In the first two cases (figure 7a and b), no instabilities were observed while in the third and fourth cases, instabilities occur. Thus low static tension can give lower loads to the shaft bearing up to some point, but excessively low static tension can cause instabilities due to the combined effect of both parametric and external excitations.

The external periodic loads were observed to change the excitation frequency and the magnitude. The common effect for all cases considered here is that the addition of external periodic loads reduces the vibration amplitude around resonances and increases the amplitude away from resonances. The effect of different excitation frequencies is shown in figure 8. The frequency changes from $4\omega_1$ to ω_1 and the effect of external load increases as the frequency decreases. Also, the effects of external load increase as the amplitude of the external load increases.

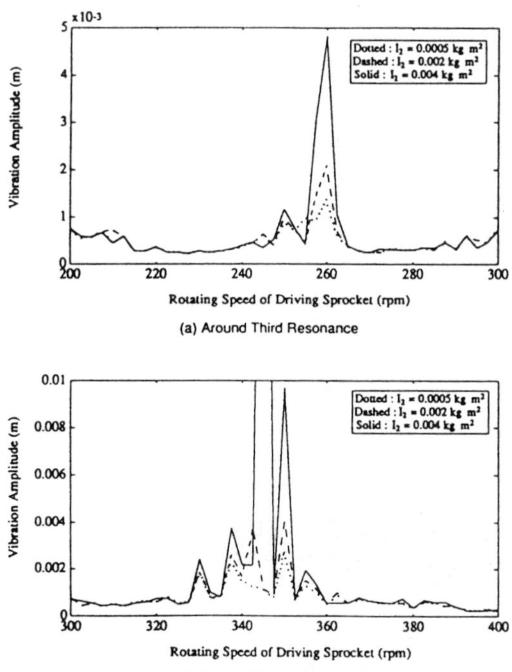

Figure 6 Effect of Moment of Inertia for Driven Sprocket System on Chain Vibration (Chain Drive : $n_1 = n_2 = 24$, cd = 20.3 pitches)

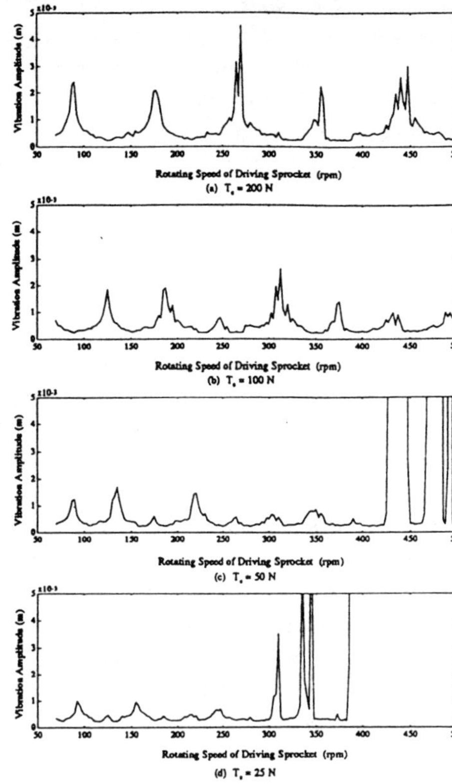

Figure 7 Effect of Static Tension on Chain Vibration (Chain Dirve : $n_1 = n_2 = 24$, cd = 19.7 pitches)

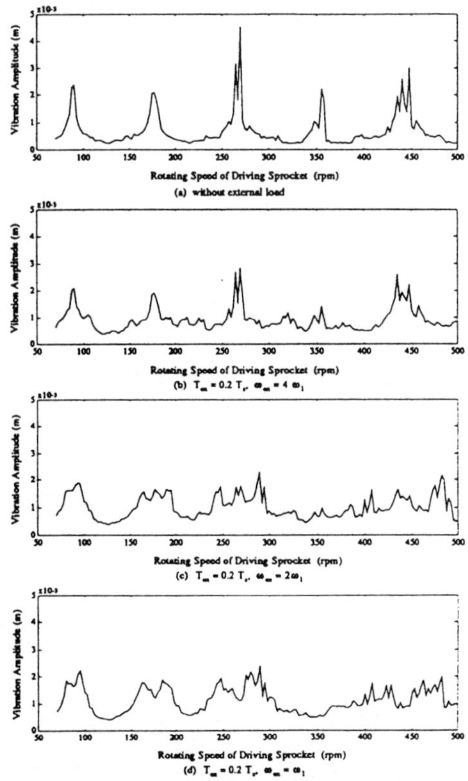

Figure 8 Effect of Periodic External Load Frequency on Chain Vibration (Chain Dirve : $n_1 = n_2 = 24$, cd = 19.7 pitches)

The motion of a chain drive at medium operating speeds is characterized by the dynamic load due to polygonal action based on the elastic four bar mechanism. There are two possible instabilities due to the dynamic load. The existence of the instabilities is limited by the operating speed, the inverse ratio of the sprocket tooth numbers, and the moment of inertia for the driven sprocket system. Figure 9 shows both the cases with and without the instabilities. The possibility of the instabilities increases as the ratio becomes larger and the moment of inertia becomes smaller.

The behavior of the chain drive near the first longitudinal resonance is observed in figure 10 which was produced by solving both equations of motion for transverse and longitudinal vibrations simultaneously. The vibration amplitudes near the first longitudinal resonance and near the fifth transverse resonance (located a little above the longitudinal resonance) were observed. The vibration amplitude is only weakly amplified near the first resonance because the magnitude of the dynamic load due to polygonal action is very small and the fundamental component of the dynamic load due to impact is not very large. However, the coupling between the longitudinal vibration and the transverse vibration is very clear near the transverse resonance. Accordingly it is suggested that operation near the transverse resonance close to the first longitudinal resonance should be avoided.

4. SUMMARY AND CONCLUSION

The chain drive model presented here includes several important features such as impact phenomenon, polygonal action, periodic length change, external periodic loads and nonlinear coupling between transverse and longitudinal motions at high operating speeds. At low operating speeds, the polygonal action and the external periodic loads govern the vibration characteristics largely through the transverse displacement excitations and the periodic tension variations. For a given center distance, multiple peaks can occur in the vicinity of each linear resonance due to the periodic length change. The vibration responses are affected by center distances over a wide range and the instabilities around the linear resonances can occur due to the periodic tension variation. The external periodic load with a frequency of the rotating speed of the driving sprocket can reduce the amplitudes around the resonances and increase the amplitude away from the resonances. At medium operating speeds, there is a danger of instabilities due to the two resonances of the dynamic load due to polygonal action. But the instabilities do not always occur. The possibility of instabilities

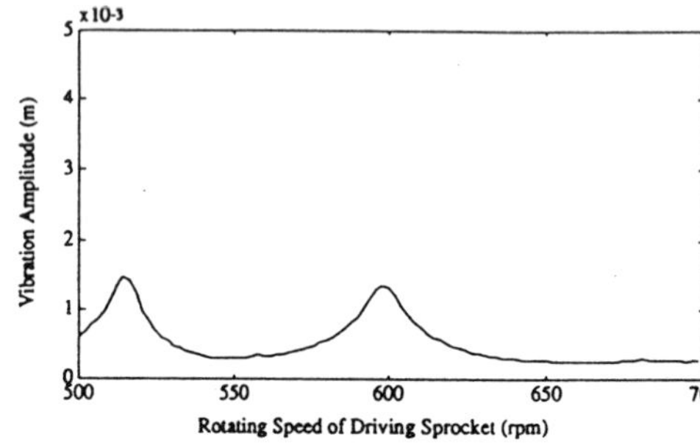

(a) No instabilities due to polygonal action in elastic chain span
(n_1 = 24, n_2 = 24, I_2 = 0.002)

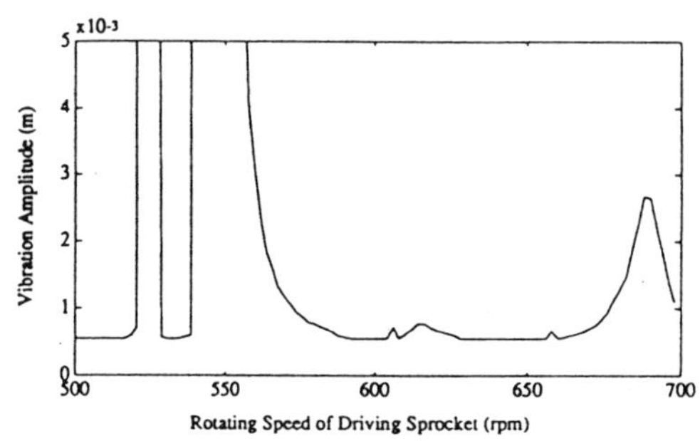

(b) Instabilities due to polygonal action in elastic chain span
(n_1 = 15, n_2 = 40, I_2 = 0.0005)

Figure 9 Vibraion of Chain Drive at Medium Operating Speeds (Chain Drive : cd = 19.7 pitches)

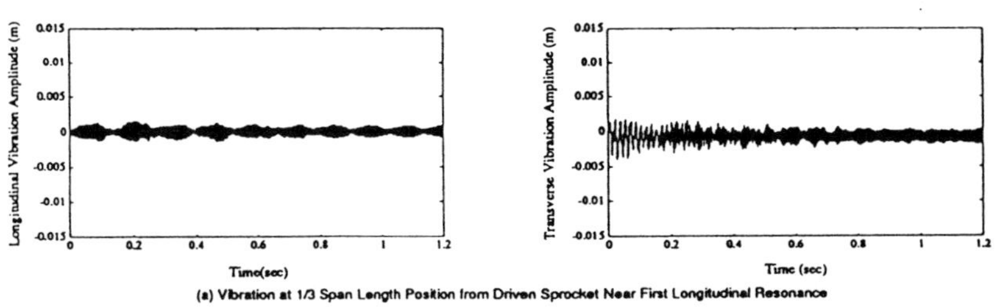

(a) Vibration at 1/3 Span Length Position from Driven Sprocket Near First Longitudinal Resonance

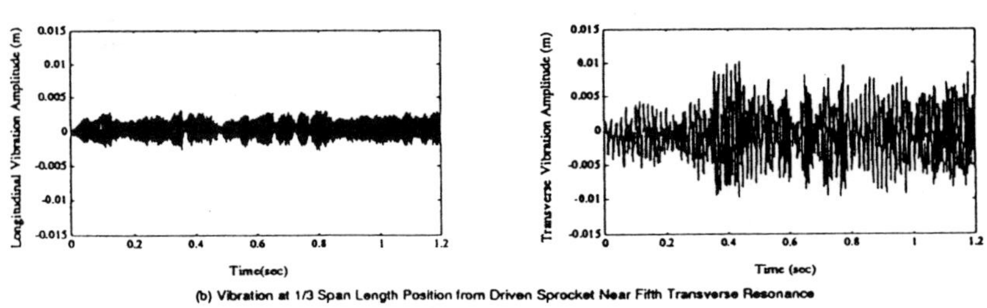

(b) Vibration at 1/3 Span Length Position from Driven Sprocket Near Fifth Transverse Resonance

Figure 10 Vibration Signals Changing from Translent to Steady State at High Operating Speeds (Chain Drive : n_1 = n_2 = 24, cd = 20.3 pitches)

increases as the ratio of n_2/n_1 becomes larger and the moment of inertia for the driven sprocket system becomes smaller. At high operating speeds, the coupling between transverse and longitudinal vibrations is not significant around the first longitudinal vibration because of the small magnitude of the excitation but the coupling between them is clear around the transverse resonance slightly above the longitudinal resonance.

The suggested model still has its limitations and in order to investigate the vibration characteristics of roller chain drives more exactly, refined models of impact and polygonal action interacting with span vibration should be developed. Experimental work will be necessary to augment the theoretical work.

5. ACKNOWLEDGEMENTS

The authors gratefully acknowledge the support of the National Science Foundation through Grants No. MSM-88-12957 and MSS-8996293. We also wish to thank N. C. Perkins for his thoughtful suggestions during the early stage of the model development.

BIBLIOGRAPHY

Ariaratnam, S. T. and Asokanthan, S. F., 1987, "Dynamic Stability of Chain Drives", ASME *Journal of Mechanism, Transmission, and Automation in Design*, vol. 109, no. 3, pp. 412 - 418.

Binder, R. C., 1956, "Mechanics of the Roller Chain Drive", Prentice-Hall, 204 p.

Bouillon, G. and Tordian, G. V., 1965, "On Polygonal Action in Roller Chain Drives," ASME *Transactions*, vol. 87, pp. 243 - 250.

Chew, M., 1985, "Inertia Effects of a Roller Chain on Impact Intensity", ASME *Journal of Mechanism, Transmission, and Automation in Design*, vol. 107, pp. 123-130.

Chen,C. K. and Freudenstein, F., 1988, "Toward a More Exact Kinematics of Roller Chain Drives", *ASME Journal of Mechanism, Transmission, and Automation in Design*, vol. 110, pp.269-275.

Conwell, J. C., Johnson, G. E., and Peterson, S. W., 1992a, "Design and Construction of a Machine to Evaluate the Forces in Roller Chain Drives", Proceedings of the 1992 International Power Transmission and Gearing Conference, DE-vol. 43-2, ASME, pp.703-709.

Conwell, J. C., Johnson, G. E., and Peterson, S. W., 1992b, "Experimental Investigation of the Forces in a Link Side Plate During Normal Operation of a Roller Chain Drive", Proceedings of the 1992 International Power Transmission and Gearing Conference, DE-vol. 43-2, ASME, pp.711-716.

Conwell, J. C., Johnson, G. E., and Peterson, S. W., 1992c, "Experimental Investigation of the Impact Force that Occurs When a Roller Seats on the Sprocket During Normal Operation of a Roller Chain Drive", Proceedings of the 1992 International Power Transmission and Gearing Conference, DE-vol. 43-2, ASME, pp.717-721.

Dubowsky, S. and Freudenstein, F., 1971, "Dynamic Analysis of Mechanical Systems with Clearances : Part 1 (Formulation of Dynamic Model)", ASME *Journal of Engineering for Industry*, pp. 305 - 309.

Fawcett, J. W. and Nicol, S. W., 1979, " A Theoretical Investigation of the Vibration of Roller Chain Drives", *Proc. Fifth World Congress on the Theory of Machines and Mechanisms*, pp. 1482 - 1485.

Jackson and Moreland, 1955, "Design Manual for Roller and Silent Chain Drives", Association of Roller and Silent Chain Manufactures, 95 pp.

Kim, M., 1990, "Dynamic Behavior of Roller Chain Drives at Moderate and High Speeds", Ph. D. dissertation, The University of Michigan, 122 p.

Kim, M. and Johnson, G. E., 1992a, "Mechanics of Roller Chain-Sprocket Contact : A General Modeling Strategy", *Proceedings of the 1992 International Power Transmission and Gearing Conference*, DE-vol. 43-2, ASME, pp. 689-695.

Kim, M. and Johnson, G. E., 1992b, "Mechanics of Roller Chain-Sprocket Contact : Observations about the Contact Phenomena and Load Distribution", *Proceedings of the 1992 International Power Transmission and Gearing Conference*, DE-vol. 43-2, ASME, pp. 697-702.

Mahalingam, S., 1957, "Transverse Vibrations of Power Transmission Chains," *British Journal of Applied Physics*, vol. 8, pp. 145 - 148.

Mahalingam, S., 1958, "Polygonal Action in Chain Drives", *Journal of Franklin Institute*, vol. 265, no. 1, pp. 23 - 28.

Morrison, R. A., 1952, "Polygonal Action in Chain Drives," *Machine Design*, vol. 24, no. 9, pp. 155 - 159.

Mote, C. D. Jr., 1965, "Some Dynamic Characteristics of Band Saws", *Forest Products Journal*, pp. 37-41.

Naguleswaran, S. and Williams, C. J. H., 1968, "Lateral Vibrations of Bandsaw Blades, Pulley Belts, and the like," *International Journal of Mechanical Sciences*, vol. 10, pp. 239 - 250.

Perkins, N. C. and Mote, C. D. Jr., 1987, "Three-dimensional Vibration of Traveling Elastic Cables", *Journal of Sound and Vibration*, vol. 114, no. 2, pp. 325 - 340.

Ryabov, G. K., 1968, "Inertia Effects of Impact Loading in Chain Drive", *Russian Engineering Journal*, vol. 48, No. 8, pp. 17-19.

Shimizu, H. and Sueoka, A., 1975, "Nonlinear Forced Vibration of Roller Chain," *Bulletin of JSME*, vol. 18, no. 124, pp. 1090 - 1100.

Turnbull, S. R. and Fawcett, J. N., 1973, "Dynamic Behavior of Roller Chain Drives," *Instn. Mech. Engrs.*, pp. 29 - 35.

Turnbull, S. R. and Fawcett , J. N., 1975, "An Approximate Kinematic Analysis of the Roller Chain Drive", *Proc. Fourth World Congress on the Theory of Machines and Mechanisms*, paper no. 168.

Tsujioka, Y., 1976, " Transverse Vibration of Roller Chain", *Trans. JSME*, vol. 42, no. 356, pp. 1087 - 1095.

Ulsoy, A. G., Whitesell, J. E. and Hooven, M. D., 1985 "Design of Belt-Tensioner Systems for Dynamic Stability", ASME *Journal of Vibration, Stress, and Reliability in Design*, vol. 107, pp. 282 - 290.

Veikos, N. C. and Freudenstein, F., 1992, "On the Dynamic Analysis of Roller Chain Drives : Part 1- Theory, Part 2- Case Study", *Mechanism Design and Synthesis*, DE-vol. 46, edited by Kinzel et al., ASME, NY, pp. 431-450.

Wang, K. W., 1992, "On the Stability of Chain Drive Systems Under Periodic Sprocket Oscillations", ASME *Journal of Vibration and Acoustics*, vol. 114, pp. 119 - 126.

Wang, K. W., Liu, S. P., Hayek, S. I. , and Chen, F. H. K., 1992, "On the impact intensity of Vibrating Axially Moving Roller Chains", ASME *Journal of Vibration and Acoustics*, vol. 114, pp. 397 - 403.

APPENDIX I
DYNAMIC LOAD DUE TO IMPACT

The component of contact force in the chain span direction is the dynamic load due to the impact and it is calculated assuming that the sprocket is stationary during the impact since the inertia of the sprocket is large comparing with m_i. When the roller is in contact with the tooth, its equation of motion is as follows (Fawcett and Nicol, 1979).

$$m_i \ddot{x}_0 + c_i \dot{x}_0 + k_i x_0 = 0$$

Initial conditions : $x_0(0) = 0$, $\dot{x}_0(0) = V_{rel} \sin\eta$

Then the solution is given by

$$x_0 = \frac{e^{-\zeta\omega_n t}}{\omega_d} V_{rel} \sin\eta \sin\omega_d t$$

where $\zeta = \frac{c_i}{2\sqrt{m_i k_i}}$, $\omega_n = \sqrt{\frac{k_i}{m_i}}$, $\omega_d = \omega_n\sqrt{1-\zeta^2}$.

And the solution applies until such time that $x_0 = 0$ again.

$$\sin\omega_d t_i = 0 \ , \ t_i = \frac{\pi}{\omega_d}$$

Let the dynamic load due to the impact be D_{imp}.

$$D_{imp} = (c_i\dot{x}_0 + k_i x_0)\cos\eta = \frac{V_{rel}\sin 2\eta}{2\omega_d} e^{-\zeta\omega_n t}\{(k_i - c_i\zeta\omega_n)\sin\omega_d t + c_i\omega_d\cos\omega_d t\}$$

where $0 < t < t_i$.

APPENDIX II
ELEMENTS OF MATRICES IN EQUATION (2-20) :

$a_{11} = \frac{c_1 k \omega_1}{2 N p}$, $a_{12} = \frac{a_0 k \omega_1}{p} - \frac{b_1 k \omega_1}{2 N p}$,
$a_{13} = \frac{c_2 k \omega_1}{4 N p}$, $a_{14} = \frac{a_0 k \omega_1}{2 p} - \frac{b_2 k \omega_1}{4 N p}$,
$a_{21} = \frac{c_2 k \omega_1}{2 N p} - \frac{c_2 l_2 p \omega_1}{2 a_0^2 N^3}$,
$a_{22} = -\frac{a_0 k \omega_1}{p} + \frac{b_1 k \omega_1}{N p} - \frac{b_2 k \omega_1}{2 N p} + \frac{b_2 l_2 p \omega_1}{2 a_0^2 N^3} + \frac{l_2 p \omega_1}{a_0 N^2}$,
$a_{23} = \frac{c_1 k \omega_1}{4 N p} + \frac{c_1 l_2 p \omega_1}{2 a_0^2 N^3}$, $a_{24} = \frac{b_1 k \omega_1}{4 N p} - \frac{b_1 l_2 p \omega_1}{2 a_0^2 N^3}$,
$a_{31} = -\frac{c_1 k \omega_1}{2 N p} - \frac{c_1 l_2 p \omega_1}{a_0^2 N^3}$, $a_{32} = -\frac{b_1 k \omega_1}{2 N p} + \frac{b_2 k \omega_1}{N p}$
$- \frac{b_1 l_2 p \omega_1}{a_0^2 N^3}$, $a_{33} = 0$, $a_{34} = -\frac{a_0 k \omega_1}{2 p} + \frac{b_2 k \omega_1}{2 N p} + \frac{2 l_2 p \omega_1}{a_0 N^2}$,
$a_{41} = -\frac{c_2 k \omega_1}{2 N p} - \frac{3 c_2 l_2 p \omega_1}{2 a_0^2 N^3}$, $a_{42} = -\frac{b_2 k \omega_1}{2 N p} - \frac{3 b_2 l_2 p \omega_1}{a_0^2 N^3}$,
$a_{43} = -\frac{c_1 k \omega_1}{4 N p} - \frac{3 c_1 l_2 p \omega_1}{2 a_0^2 N^3}$, $a_{44} = -\frac{b_1 k \omega_1}{4 N p} - \frac{3 b_1 l_2 p \omega_1}{2 a_0^2 N^3}$,
$a_{51} = a_{52} = 0$,
$a_{53} = -\frac{c_2 k \omega_1}{4 N p} - \frac{2 c_2 l_2 p \omega_1}{a_0^2 N^3}$, $a_{54} = -\frac{b_2 k \omega_1}{4 N p} - \frac{2 b_2 l_2 p \omega_1}{a_0^2 N^3}$,
$a_{61} = \frac{a_0 k \omega_1}{p} - \frac{b_2 k \omega_1}{2 N p} + \frac{b_2 l_2 p \omega_1}{2 a_0^2 N^3} - \frac{l_2 p \omega_1}{a_0 N^2}$, $a_{62} = \frac{c_1 k \omega_1}{N p}$
$- \frac{c_2 k \omega_1}{2 N p} + \frac{c_2 l_2 p \omega_1}{2 a_0^2 N^3}$, $a_{63} = \frac{b_1 k \omega_1}{4 N p} + \frac{b_1 l_2 p \omega_1}{2 a_0^2 N^3}$,
$a_{64} = \frac{3 c_1 k \omega_1}{4 N p} + \frac{c_1 l_2 p \omega_1}{2 a_0^2 N^3}$, $a_{71} = \frac{b_1 k \omega_1}{2 N p} + \frac{b_1 l_2 p \omega_1}{a_0^2 N^3}$,
$a_{72} = -\frac{c_1 k \omega_1}{2 N p} + \frac{c_2 k \omega_1}{N p} - \frac{c_1 l_2 p \omega_1}{a_0^2 N^3}$, $a_{73} = \frac{a_0 k \omega_1}{2 p} - \frac{2 l_2 p \omega_1}{a_0 N^2}$,
$a_{74} = \frac{c_2 k \omega_1}{2 N p}$, $a_{81} = \frac{b_2 k \omega_1}{2 N p} + \frac{3 b_2 l_2 p \omega_1}{2 a_0^2 N^3}$,
$a_{82} = -\frac{c_2 k \omega_1}{2 N p} - \frac{3 c_2 l_2 p \omega_1}{2 a_0^2 N^3}$, $a_{83} = \frac{b_1 k \omega_1}{4 N p} + \frac{3 b_1 l_2 p \omega_1}{2 a_0^2 N^3}$,
$a_{84} = -\frac{c_1 k \omega_1}{4 N p} - \frac{3 c_1 l_2 p \omega_1}{2 a_0^2 N^3}$, $a_{91} = a_{92} = 0$, $a_{93} = \frac{b_2 k \omega_1}{4 N p}$
$+ \frac{2 b_2 l_2 p \omega_1}{a_0^2 N^3}$, $a_{94} = -\frac{c_2 k \omega_1}{4 N p} - \frac{2 c_2 l_2 p \omega_1}{a_0^2 N^3}$, $x_1 = l_1$,
$x_2 = m_1$, $x_3 = l_2$, $x_4 = m_2$,
$f_1 = \frac{a_1 c_1 k \omega_1}{2 p} + \frac{a_2 c_2 k \omega_1}{4 p}$, $f_2 = \frac{a_2 c_1 k \omega_1}{4 p} + \frac{a_1 c_2 k \omega_1}{2 p}$
$+ \frac{c_1 l_2 p \omega_1}{a_0 N^2}$, $f_3 = -\frac{a_1 c_1 k \omega_1}{2 p} + \frac{2 c_2 l_2 p \omega_1}{a_0 N^2}$, $f_4 = -\frac{a_2 c_1 k \omega_1}{4 p}$
$- \frac{a_1 c_2 k \omega_1}{2 p}$, $f_5 = -\frac{a_2 c_2 k \omega_1}{4 p}$, $f_6 = \frac{a_2 b_1 k \omega_1}{4 p} - \frac{a_1 b_2 k \omega_1}{2 p}$
$+ \frac{a_0 a_1 k N \omega_1}{p} - \frac{b_1 l_2 p \omega_1}{a_0 N^2}$, $f_8 = \frac{a_2 b_1 k \omega_1}{4 p} + \frac{a_1 b_2 k \omega_1}{2 p}$,
$f_7 = \frac{a_1 b_1 k \omega_1}{2 p} + \frac{a_0 a_2 k N \omega_1}{2 p} - \frac{2 b_2 l_2 p \omega_1}{a_0 N^2}$, $f_9 = \frac{a_2 b_2 k \omega_1}{4 p}$.

APPENDIX III
COEFFICIENTS IN EQUATION (2-17) :

$q0 = -0.5\,(b_1 l_1 + b_2 l_2 + c_1 m_1 + c_2 m_2) + (a_0 N)^2$
$q1 = \frac{1}{p}\{-0.5\,(b_2 l_1 + b_1 l_2 + c_2 m_1 + c_1 m_2) - a_0 n\,(b_1 - l_1)\}$
$q2 = \frac{1}{2p}\{-0.5\,(b_1 l_1 - c_1 m_1) - a_0 N\,(b_2 - l_2)\}$,
$q3 = \frac{1}{3p}\{-0.5\,(b_2 l_1 + b_1 l_2 - c_2 m_1 - c_1 m_2)\}$,
$q4 = \frac{1}{4p}\{-0.5\,(b_2 l_2 - c_2 m_2)\}$
$p1 = \frac{1}{p}\{-0.5\,(c_2 l_1 - c_1 l_2 - b_2 m_1 + b_1 m_2) - a_0 N\,(c_1 - m_1)\}$
$p_2 = \frac{1}{2p}\{-0.5\,(c_1 l_1 + b_1 m_1) - a_0 N\,(c_2 - m_2)\}$
$p_3 = \frac{1}{3p}\{-0.5\,(c_2 l_1 + c_1 l_2 + b_2 m_1 + b_1 m_2)\}$,
$p4 = \frac{1}{4p}\{-0.5\,(c_2 l_2 + b_2 m_2)\}$

DE-Vol. 63, Vibrations of Mechanical Systems
and the History of Mechanical Design
ASME 1993

DYNAMICS OF CURVED BEAMS UNDERGOING LARGE OVERALL MOTIONS USING THE MODE DECOMPOSITION CONCEPT

Joseph F. D'Costa
Dept. of Mechanical Engineering
University of Minnesota
Minneapolis, Minnesota

Henryk K. Stolarski
Dept. of Civil and
Mineral Engineering
University of Minnesota
Minneapolis, Minnesota

Arthur G. Erdman
Dept. of Mechanical Engineering
University of Minnesota
Minneapolis, Minnesota

ABSTRACT

A fully nonlinear formulation for the dynamics of initially curved and twisted beams, undergoing arbitrary spatial motions, is presented. The formulation admits finite bending, shearing and extension of the beam. The Mode decomposition method is employed to modify the strains in the finite element discretization process leading to the elimination of shear and membrane locking phenomena that arise in curved elements. The model incorporates all inertia effects and is capable of accurately capturing the phenomena of dynamic stiffening due to the coupling of the axial and membrane forces to the flexural deformation. All motion is referred to the inertial frame. The nonlinear formulation is suitable for modeling flexible multibody systems. Examples are presented to illustrate the validity of the proposed formulation.

1. INTRODUCTION

The dynamics of multibody systems is of great importance in the design and analysis of mechanisms, robots, spacecraft, modern high-speed precision machinery and bio-mechanical systems. In its most general form a multibody system can be described as an interconnected assembly of rigid and deformable bodies that can undergo large overall motion. Classically the dynamics of such systems has been studied by assuming that the component bodies are rigid.

Modern technological advances have dictated that the rigid body assumption be dropped. This is specially true in the case of high-speed, light-weight systems where flexibility effects can significantly influence dynamic behavior (Liou and Erdman, 1987). For instance, the deformation of an elastic body undergoing large overall motion will couple with the rigid body motion of the body through nonlinear kinematics. In rotating beam and plate structures, this effect manifests itself as centrifugal stiffening (Kane et al. 1987) and (Simo and Vu-Quoc, 1987) – causing the tip displacement of a spinning beam *not* to diverge as the spin rate approaches the natural frequency of the first bending mode, contrary to the predictions of a beam model based upon a linearized strain measure.

Multibody systems, more often than not, incorporate curved structural members as component bodies. In mechanism design, for example, it is commonplace to find that a *curved beam* is required to satisfy constraints in the workspace. Initial twist and curvature, introduced either by design or as manufacturing defects, are characteristic of practical multibody systems. An experimental and analytical study (Karkoub, 1990) of the dynamic analysis of a high-speed four-bar planar mechanism, revealed that at high speeds torsional effects and out of plane motion come into play in such systems. They have demonstrated that these effects alter the natural frequency of the system significantly. It is a well known fact that an initial curvature of a link of a mechanism can induce larger longitudinal deformantion and a greater internal moment as compared to a straight link. Thus, the modeling of curved structural members is a necessity for the accurate modeling of multibody system dynamic behavior. A general beam model formulated for multibody-dynamics should be capable of accurately representing deformation of the beam due to extension, bending, transverse shear, torsion and inertial coupling. Such beam theories, especially in the context of multibody-dynamics, have been developed (Kane, et al. 1987), (Boutaghou, et al. 1992), (Simo, 1985) and (Downer, et al. 1990). The effects of shear deformation and rotary inertia on the nonlinear dynamics of rotating curved beams has been studied in (Bartolone and Shabana, 1989). However, a general formulation should also be able to represent arbitrary geometries.

The finite element method is used almost exclusively in the analysis of the flexible multibody systems. There are however serious difficulties associated with the finite element modeling of curved structural members. These are related to the fact that in these structures, an inappropriate formulation can cause an artificial constraint to appear whereby bending-dominated response is replaced by a membrane-dominated response (Stolarski and Belytschko, 1982). A novel formulation for beam and shell elements (Wang, 1989), attempts to eliminate "locking" using a kinematically based approach. This is in contrast to the reduced integration, mixed approach, assumed strain approach and other approaches traditionally employed in solving the problem.

The approach discussed here generalizes the above approach to the dynamical case. The formulation under discussion is a fully nonlinear three-dimensional formulation valid for curved beams and includes finite extension, finite shearing of the rod, and torsion. Besides eliminating the undesirable locking mechanism, the formulation is physically intuitive.

2. KINEMATICS OF DEFORMATION

The model presented here is developed with the following assumptions: 1. A cross-section which is originally plane and orthogonal to the line of centroids of the beam, remains plane but not necessarily orthognal to the centroidal axis, after the deformation. 2. Deformations in the plane of the cross-section are negligible.

All motion is referred to fixed Cartesian frame with the unit vectors $\vec{e}^{k} = \vec{e}_k$, k=1,2,3. The beam is parametrized by a system of curvilinear coordinates ξ^K, $K = 1,2,3$, affixed to the material particles. These coordinates identify the particles during the entire deformation process. The parameter ξ^3 parametrizes the length along the centroidal axis and the parameters ξ^1 and ξ^2 parametrize the cross-section. If X^k and x^k denote the Cartesian coordinates of the particles in the beam in the reference and the deformed configuration, respectively, then the position vector of any particle in the beam can be expressed as follows: $\hat{\vec{R}} = X^k(\xi^L)\vec{e}_k$ (original configuration); and $\hat{\vec{r}} = x^k(\xi^L)\vec{e}_k$ (deformed configuration).

In addition, an orthonormal triad of vectors, $(\vec{t}_1(\xi^3), \vec{t}_2(\xi^3), \vec{t}_3(\xi^3))$, is rigidly attached at the centroid of each cross-section with $\vec{t}_3$ being normal to the cross-section. As a result the position vector of a typical point in the volume of the beam can be described as (Fig. 1)

$$\hat{\vec{r}}(\xi^K, T) = \vec{r}(\xi^3, T) + \vec{t}_\Lambda(\xi^3, T)\xi^\Lambda \tag{1}$$

where ξ^Λ, $\Lambda = 1,2$; $\vec{r}$ is the position vector of the centroidal axis, and T, is the time parameter. The vector $\vec{t}_\Lambda(\xi^3)$ are assumed to satisfy the following conditions:

$$\vec{t}_\Lambda \cdot \vec{t}_\beta = \delta_{\Lambda\beta}, \quad \Lambda, \beta = 1,2$$

where $\delta_{\Lambda\beta}$ is the Kronecker delta.

With the above kinematic assumptions the vectors of the natural basis, at any point of the beam, are:

$$\hat{\vec{g}}_K \equiv \hat{\vec{r}}_{,K} = (\vec{r}_{,3} + \vec{t}_{\Lambda,3}\xi^\Lambda)\delta^3{}_K + \vec{t}_\Lambda \delta^\Lambda{}_K \tag{2}$$

$$\equiv \vec{g}_K + \vec{t}_{\Lambda,3}\xi^\Lambda \delta^3{}_K \tag{3}$$

where $\vec{g}_K = \hat{\vec{g}}_K|_{\xi^3=0}$ is the natural basis evaluated at the points of the centroidal axis. The comma represents differentiation with respect to the coordinate ξ^K. Dual basis $\vec{g}^K$ is defined through the orthogonality condition

$$\vec{g}_K \cdot \vec{g}^L = \delta_K{}^L$$

The above formulas are valid in any configuration. To identify the original configuration, however, we use uppercase letters. Thus in the original configuration we have $\vec{r} \to \vec{R}, \vec{t}_K \to \vec{T}_K$, $\vec{g}_K \to \vec{G}_K$. (Note that Eqs. (1) and (3) imply $\vec{t}_\Lambda = \vec{g}_\Lambda$), $\vec{g}^K = \vec{G}^K$. In addition we assume that $\vec{T}_3 = \vec{G}_3 = \vec{R}_{,3}$ which reflects the fact that in the original configuration the cross-section is orthogonal to the centroidal axis. Figure 1 shows the configurations of the beam and the associated deformation kinematics. (The significance of the inextensional configuration will be explained in section 5).

Since the vectors constituting the triad remain orthonormal during the deformation process in accordance with our assumptions, the orientation of the triad, valid even for finite rotations, is given by an orthogonal rotation matrix Q such that:

$$\vec{t}_\Lambda = Q\vec{T}_\Lambda \tag{4}$$

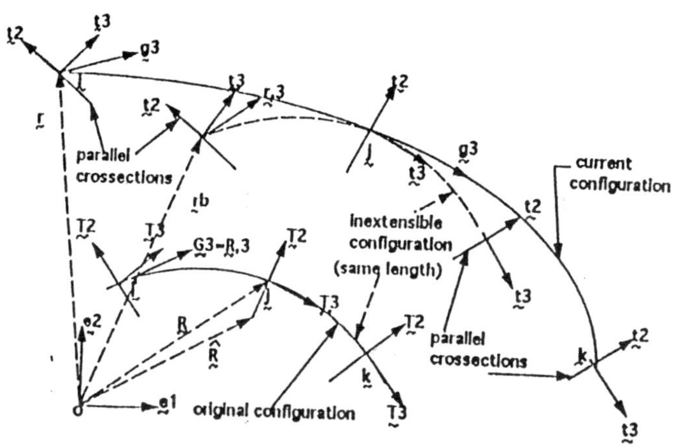

Figure 1: THE CONFIGURATIONS OF THE BEAM

The inertia terms evaluated in the sequel involve the derivatives of the vectors constituting the triad $\vec{t_K}$, $K = 1, 2, 3$. Beginning with Eq. (4) we note that

$$\dot{\vec{t_\Lambda}} = \vec{w} \times \vec{t_\Lambda} \tag{5}$$

$$\begin{aligned} \ddot{\vec{t_\Lambda}} &= \dot{\vec{w}} \times \vec{t_\Lambda} + \vec{w} \times \dot{\vec{t_\Lambda}} \\ &= \dot{\vec{w}} \times \vec{t_\Lambda} + \vec{w} \times (\vec{w} \times \vec{t_\Lambda}) \end{aligned} \tag{6}$$

where $\vec{w}$ is the axial vector of the skew-symmetric tensor $\dot{Q}Q^T$. The variations $\delta\vec{t_\Lambda}$ which are part of $\delta\vec{r}$ (to be used in the variational statement of the problem) are analogical to the velocities $\dot{\vec{t_\Lambda}}$ specified in Eq.(5)

$$\delta\vec{t_\Lambda} = \delta\vec{\omega} \times \vec{t} \tag{7}$$

where $\delta\vec{\omega}$ is the axial vector of the skew-symmetric tensor $\delta Q Q^T$.

3. VARIATIONAL FORMULATION OF THE EQUATIONS OF MOTION

3.1 Generalized Strain Measures

Given the above kinematics of deformation, the Lagrangian formulation for the dynamics of the beam is constructed. The Green-Lagrange strain tensor E is used as a measure of strain,

$$E = \frac{1}{2}(F^T F - I) \tag{8}$$

where $F = \vec{g}_K \otimes \vec{G}^K$ is the deformation gradient tensor, The covariant curvilinear strain components of the Green-Lagrange strain tensor can be written using Eq.(1) and (3) in the following form: (Wang and Stolarski, -b)

$$E_{KL} = \frac{1}{2}(\hat{\vec{r}}_{,K} \cdot \hat{\vec{r}}_{,L} - \hat{\vec{R}}_{,K} \cdot \hat{\vec{R}}_{,L}) \tag{9}$$

The non-zero components of Eq.(9),can be written in the following way

$$\begin{aligned} E_{33} &= \epsilon + \xi^\Lambda \kappa_\Lambda && (10) \\ 2E_{13} &= \gamma_1 - \xi^2 \tau && (11) \\ 2E_{23} &= \gamma_2 + \xi^1 \tau && (12) \end{aligned}$$

where

$$\begin{aligned} \epsilon &= \tfrac{1}{2}(\vec{r}_{,3} \cdot \vec{r}_{,3} - \vec{R}_{,3} \cdot \vec{R}_{,3}) && (13) \\ \kappa &= \vec{r}_{,3} \cdot \vec{t}_{\Lambda,3} - \vec{R}_{,3} \cdot \vec{T}_{\Lambda,3} \quad \Lambda = 1,2 && (14) \\ \tau &= \vec{t}_{1,3} \cdot t_2 - \vec{T}_{1,3} \cdot T_2 && (15) \\ &= -\vec{t}_{2,3} \cdot t_1 + \vec{T}_{2,3} \cdot T_1 \\ \gamma_\Lambda &= \vec{r}_{,3} \cdot t_\Lambda - \vec{R}_{,3} \cdot T_\Lambda && (16) \end{aligned}$$

are the generalized strains: axial elongation, curvatures, twist and transverse shear strains respectively.

3.2 Weak Form of the Governing Equations

The weak formulation of the problem at hand reads as follows:

$$\int_\Omega \rho\ddot{\vec{r}} \cdot \delta\vec{r} + \int_\Omega S{:}\delta E d\Omega = \int_\Omega \vec{b} \cdot \delta\vec{r} d\Omega + \int_{\partial\Omega} \vec{t} \cdot \delta\vec{r} dS \tag{17}$$

where ρ is the mass density, $\ddot{\vec{r}}$ is the acceleration of the point of the beam with the position vector $\vec{r}$ in the current configuration, $\delta\vec{r}$ is the virtual displacement of the point consistent with the constraints, $\vec{b}$ represents the body forces, $\vec{t}$ is the surface traction, S is the second Piola-Kirchoff stress tensor, δE is the variation of the Green-Lagrange strain tensor, Ω is the volume of the beam and $\partial\Omega$ denotes the boundary (surface) of the beam.

In the light of Eq. (1), (5), and (7), the first term in Eq. (17) becomes

$$\int_\Omega \rho\ddot{\vec{r}} \cdot \delta\vec{r} = \tag{18}$$

$$\int_{\Omega}(\vec{\ddot{r}}\delta\vec{r}+(\vec{\dot{w}}\times\vec{t_j}\xi^j)\delta\vec{t_k}\xi^k+(\vec{w}\times(\vec{w}\times\vec{t_j}\xi^j))\delta\vec{t_k}\xi^k)d\Omega \quad (19)$$

The second term in Eq.(17) leads to the following expression

$$\int_{\Omega}\rho\boldsymbol{S}:\delta\boldsymbol{E}d\Omega=\int_{l}(N\delta\epsilon+M^{\Lambda}\delta\kappa_{\Lambda}+T\delta\tau+S^{\Lambda}\delta\gamma_{\Lambda})dl \quad (20)$$

where l represents the arc length of the beam measured along the centroidal axis and the generalized, congugate stresses are defined as follows

$$N = \int_A S^{33}\hat{\jmath}dA \quad (21)$$
$$M^{\Lambda} = \int_A S^{33}\xi^{\Lambda}\hat{\jmath}dA \quad (22)$$
$$T = \int_A (S^{23}\xi^1 - S^{13}\xi^2)\hat{\jmath}dA \quad (23)$$
$$S^{\Lambda} = \int_A S^{3\Lambda}\hat{\jmath}dA \quad (24)$$

where A is the area of cross-section and $\hat{\jmath} = \jmath/\sqrt{G_{33}}$ with $\jmath$ being the Jacobian of the mapping $\xi^K \rightarrow X^K$ and $G_{33} = \vec{G_3}\cdot\vec{G_3}$.

3.3 Constitutive Equations

The elastic constitutive equations used in this work are of the following form

$$\tilde{\sigma} = \boldsymbol{D}\tilde{\epsilon} \quad (25)$$

where $\tilde{\epsilon}$ contains the *physical* generalized strain components defined in Eqs. (13) - (16) and $\tilde{\sigma}$ includes the corresponding physical components of the generalized forces specified in Eqs. (21) - (24). The constitutive tensor $\boldsymbol{D}$ is chosen to be the following

$$\boldsymbol{D}=\begin{bmatrix} EA & 0 & 0 & 0 & 0 & 0\\ 0 & EI_1 & 0 & 0 & 0 & 0\\ 0 & 0 & EI_2 & 0 & 0 & 0\\ 0 & 0 & 0 & CG & 0 & 0\\ 0 & 0 & 0 & 0 & \lambda_1 GA & 0\\ 0 & 0 & 0 & 0 & 0 & \lambda_2 GA \end{bmatrix} \quad (26)$$

In the above equation, A is the area of cross-section of the beam, I_Λ are the corresponding moments of inertia, G is the shear modulus, C is a constant that accounts for the fact that any cross-section other than circular would warp under torsion, λ_Λ are coefficients that incorporate the variation of shear stress across a cross-section, whereas it is assumed to be constant by our kinematic assumptions. For a rectangular cross-section, $\lambda_\Lambda = \frac{5}{6}$ and $C = k_1(W^3)(H)$ where W is the width and H is the height of the rectangular cross-section and k_1 is a factor proportional to the magnitude of W/H (Timoshenko, 1969).

4. FINITE ELEMENT APPROXIMATION

The finite element approximation is constructed using the quadratic isoparametric element. Since one of the goals of the present formulation is the elimination of shear and membrane locking, which typically arise in curved members, the quadratic element, which is the lowest order element capable of representing initial curvature , is chosen. A formulation eliminating shear and membrane locking with quadratic elements can easily be extended to other elements of arbitrary order. It is a well known fact, however, that the problem of shear and membrane locking increases as the order of the approximation decreases.

Denoting the shape functions for an element by, $(N_I(\xi^3) \quad I = 1,2,3 \quad -1 \leq \xi^3 \leq +1)$, the finite dimensional approximation of the kinematic variables can be constructed as follows

$$\vec{r}(\xi^3) = \sum_{I=1}^{3}\vec{r_I}N_I(\xi^3) \quad (27)$$

where $\vec{r_I}$ represent the position vectors of the nodal points. In a similar manner

$$\vec{t_\Lambda}(\xi^3) = \sum_{I=1}^{3}\vec{t_{\Lambda I}}N_I(\xi^3) \quad (28)$$

The variations of both $\vec{r}$ and $\vec{t_\Lambda}$ are given by

$$\delta\vec{r}(\xi^3) = \sum_{I=1}^{3}\delta\vec{r_I}N_I(\xi^3) \quad (29)$$

$$\delta\vec{t_\Lambda}(\xi^3) = \sum_{I=1}^{3}\delta\vec{t_{\Lambda I}}N_I(\xi^3) \quad (30)$$

Since the vectors $\vec{t_{1I}}$ and $\vec{t_{2I}}$ have to remain unit and orthogonal, the variation of these vectors at each node takes the form (cf. Eq. 7)

$$\delta t_{\Lambda I} = \delta\omega_I \times t_{\Lambda I} \equiv -[\check{t_{\Lambda I}}]\delta\omega_I \quad (31)$$

where the symbol $[\check{t_{\Lambda I}}]$ represents a skew-symmetric matrix whose axial vector is $t_{\Lambda I}$. Strain variations can now be obtained in an analagous manner using Eqs.(13 thru 16). The vector of element degrees of freedom has the form

$$\delta\vec{d} = [\delta\vec{d_1}^T, \delta\vec{d_2}^T, \delta\vec{d_3}^T]^T \quad (32)$$

with

$$\delta\vec{d_I} = [\delta\vec{r_{1I}}, \delta\vec{r_{2I}}, \delta\vec{r_{3I}}, \delta\vec{\omega_{1I}}, \delta\vec{\omega_{2I}}, \delta\vec{\omega_{3I}}] \quad (33)$$

Expressing the strain variations in matrix form one gets

$$\begin{Bmatrix} \delta\vec{\epsilon_1} \\ \delta\vec{\kappa_1} \\ \delta\vec{\kappa_2} \\ \delta\vec{\tau} \\ \delta\vec{\gamma_1} \\ \delta\vec{\gamma_2} \end{Bmatrix} = \begin{bmatrix} B_1, & B_2, & B_3 \end{bmatrix} \begin{Bmatrix} \delta\vec{d_1}^T \\ \delta\vec{d_2}^T \\ \delta\vec{d_3}^T \end{Bmatrix} \quad (34)$$

or

$$\delta\vec{\epsilon} = [B]\delta\vec{d} \quad (35)$$

The matrix $[B_I]$ has dimensions 6×6 and represents the influence of the degrees of freedom at node I on the strain state at any point of the element. $[B_I] =$

$$\begin{bmatrix} \frac{1}{G_{33}} N_{I,3} r^T{}_{,3} & \vec{0} \\ \frac{1}{G_{33}} N_{I,3} t^T{}_{1,3} & -\frac{1}{G_{33}} N_{I,3} r^T{}_{,3} [\breve{\vec{t_1}}] \\ \frac{1}{G_{33}} N_{I,3} t^T{}_{2,3} & -\frac{1}{G_{33}} N_{I,3} r^T{}_{,3} [\breve{\vec{t_2}}] \\ \vec{0} & -\frac{1}{\sqrt{G_{33}}} (N_{I,3} t_2{}^T [\breve{\vec{t_1}}] + N_I r^T{}_{,3} [\breve{\vec{t_2}}]) \\ \frac{1}{\sqrt{G_{33}}} N_{I,3} t_1{}^T & -\frac{1}{\sqrt{G_{33}}} N_I r^T{}_{,3} [\breve{\vec{t_1}}] \\ \frac{1}{\sqrt{G_{33}}} N_{I,3} t_2{}^T & -\frac{1}{\sqrt{G_{33}}} N_I r^T{}_{,3} [\breve{\vec{t_2}}] \end{bmatrix} \quad (36)$$

where $\vec{0}$ denotes the zero vector. Once the relationship provided in Eq. (35) is known, the internal nodal forces for a single element can be found according to Eqs. (19) and (20), which leads to

$$\underline{f}^{int} = \int_\Omega (\hat{\ddot{\vec{r}}}\delta\vec{r} + (\vec{w} \times \vec{t_j}\xi^j)\delta\vec{t_k}\xi^k$$

$$+ \int_\Omega \rho(\vec{w} \times (\vec{w} \times \vec{t_j}\xi^j)) \cdot \delta\vec{t_k}\xi^k d\Omega + \int_{l_e} B^T \sigma dl \quad (37)$$

External nodal loads can be evaluated for a distributed line load $\vec{q}(\xi^3)$ as follows

$$\underline{f}^{ext} = \int_{l_e} N^T \vec{q} dl \quad (38)$$

5. THE MODE DECOMPOSITION TECHNIQUE

The approximating process described in the preceeding section leads to locking, particularly due to the membrane effects (Stolarski and Belytschko, 1982) and (Wang and Stolarski, -a). There are several techniques to remedy this problem, among them the reduced integration method (Stolarski and Belytschko, 1982). The technique developed here is based on the observation that in order to avoid locking, inextensional bending should be given priority over the shear and membrane deformations (Wang, 1989). Therefore for any given configuration described by the position vector $\vec{r_I}$ and vector $\vec{t_{\Lambda_I}}$, an imaginary inextensional bending configuration is found. This configuration is described by the same vectors $\vec{t_{\Lambda_I}}$ and a different position vector $\vec{r}^b_I$ such that $\vec{t_\Lambda}$ and $\vec{r}^b$ give $\epsilon = \gamma_\Lambda \approx 0$. This is achieved at the nodal locations by relating the strains ϵ and γ_Λ to $\vec{t_\Lambda}$ and the difference $\vec{r} - \vec{r}^b$ rather than $\vec{r}$. Thus whenever nodal values of $\vec{r}$ coincide with those of $\vec{r}^b$ the values of ϵ and γ_Λ are zero – creating a configuration wherein bending occurs without the involvement of membrane and strain energy. The mathematical statement of the above proposition is presented next.

Let $\vec{T_{3I}}$ be a unit vector at node I such that $(\vec{T_{1I}}, \vec{T_{2I}}, \vec{T_{3I}})$ form an orthonormal basis Eq.(3) then the vector $\vec{G_{3I}} = \vec{R}_{,3I}$ can be represented as

$$\vec{G_{3I}} = G^k{}_{3I}\vec{T_{kI}} \quad k = 1,2,3 \quad no\, sum\, on\ I \quad (39)$$

where $G^k{}_3$ are the scalar components of $\vec{G_3}$. For the given current configuration, the vectors $\vec{t_{KI}}$ $\vec{t_{3I}} = \vec{t_{1I}} \times \vec{t_{2I}}$ form an orthonormal basis at node I Eqs. (1),(3). We construct the vector tangent to the midline of the imaginary inextensionally bent configuration as follows

$$\begin{aligned} \vec{r}^b_{,3I} &\equiv \vec{g}^b_{3I} \\ &\equiv G^k{}_{3I}\vec{t_{kI}} \end{aligned} \quad (40)$$

The vector $\vec{g}^b_{3I}$ has the same length as $\vec{G_{3I}}$ and the same position relative to $\vec{t_{KI}}$ as $\vec{G_{3I}}$ does relative to $\vec{T_{KI}}$. The vector $\vec{r}^b{}_{,3}$ along the centerline of the element is constructed by integrating the following equation

$$\vec{r}^b{}_{,3} = \sum_1^3 N_I \vec{g}^b_{3I} \quad (41)$$

and assuming that the resulting curve is shifted so that at the node $I = 2$ $(\xi^3 = 0)$ $\vec{r}^b(0) = \vec{r_2}$. The expression for $\vec{r}^b$ then takes the form

$$\vec{r}^b = \vec{r_2} + \sum_I (\bar{N}_I) G^k{}_{3I}\vec{t_{kI}} \quad p = 1,2,3 \quad (42)$$

where

$$\begin{aligned} \bar{N}_1 &= -\tfrac{5}{12}N_1 - \tfrac{1}{12}N_3 \\ \bar{N}_2 &= -\tfrac{2}{3}N_1 - \tfrac{2}{3}N_3 \\ \bar{N}_3 &= -\tfrac{1}{12}N_1 - \tfrac{5}{12}N_3 \end{aligned}$$

Once the inextensional bending configuration is defined the strains ϵ and γ_Λ are calculated as follows

$$\begin{aligned} \bar{\epsilon} &= \tfrac{1}{2}(\vec{r}_{,3} \cdot \vec{r}_{,3} - \vec{r}^b{}_{,3}\vec{r}^b{}_{,3}) \\ \gamma_\Lambda &= (\vec{r} - \vec{r}^b)_{,3} \cdot t_\Lambda \end{aligned} \quad (43)$$

while the bending and twisting strain components remain unmodified. The corresponding strain variations are adjusted and this is reflected in an additional matrix $[B]^M = [B_1{}^M, B_2{}^M, B_3{}^M]$. where $[B_I]^M =$

$$\begin{bmatrix} \vec{0} & -\frac{1}{G_{33}} \bar{N}_{I,3} G^K{}_{3I} (\vec{r^b}_{,3})^T [\check{\vec{t}}_{I_K}] \\ \vec{0} & \vec{0} \\ \vec{0} & \vec{0} \\ \vec{0} & \vec{0} \\ \vec{0} & -\frac{1}{\sqrt{G_{33}}} \{ \bar{N}_{I,3} G^K{}_{3I} t_1{}^T [\check{\vec{t}}_{I_K}] + \bar{N}_{I,3} \vec{r^b}_{,3}{}^T [\check{\vec{t}}_{1_I}] \} \\ \vec{0} & -\frac{1}{\sqrt{G_{33}}} \{ \bar{N}_{I,3} G^K{}_{3I} t_2{}^T [\check{\vec{t}}_{I_K}] + \bar{N}_{I,3} \vec{r^b}_{,3}{}^T [\check{\vec{t}}_{2_I}] \} \end{bmatrix} \quad (44)$$

The variation of the strain-displacement relation now takes the form

$$\delta\bar{\epsilon} = (\underline{B} - \underline{B}^M)\delta d = \bar{B}\delta d \quad (45)$$

The crux of the formulation is the strain modification introduced by Eq. (43) and(45). This modification eliminates locking and makes it possible to obtain an accurate dynamical response of the curved beam without a recourse to numerical artifice such as reduced integration and its attendant consequences.

6. LINEARIZATION

Since the equations of motion obtained from the weak form are nonlinear, a solution scheme, such as the Newton-Raphson procedure, is required for the solution of the governing differential equations. This necessiates that the weak form be linearized. In the present work, the time discretization is achieved by the use of the Newmark-β time integration algorithm (Simo and Vu-Quoc, 1986). The resulting equations are linearized to obtain the effective tangent stiffness matrix. The linearization procedure takes into consideration the strain modification discussed earlier Eq. (45). The tangent stiffness matrix obtained is thus very different from those arising from existing, standard formulations.

7. NUMERICAL RESULTS

To validate the approach presented in this work, numerical solutions to typical problems are presented. The theory presented applies both to the static and dynamic cases. To validate the approach for the static case, the problem involving a 45 deg cantilever beam under a concentrated force is solved for. This problem is depicted in (Fig. 2) The figure shows an isometric projection of the beam at different stages of the large deformation process. The beam was discretized into four elements. The original configuration of the beam is located in the x-y plane as shown in (Fig. 2) and forms one eight of the full circle of radius $R = 100$. One end of it is totally fixed while the other is free and loaded by a gradually increasing concentrated force parallel to the z axis. The force-displacement relationship is shown in (Fig. 3). It agrees very well with the results obtained by Simo, et al.(19- 86).

The results from the dynamic analysis will be presented at the conference.

8. ACKNOWLEDGEMENTS

Support of the Minnesota Supercomputer Institute in the form of a Computer time grant is gratefully acknowledged.

REFERENCES

Bartolone, D. F., and Shabana, A. A., "Effect of Beam Initial Curvature on the Dynamics of Deformable Multibody Systems", *Mechanism and Machine Theory,* 1989, vol. 24, no. 5, pp. 411 – 429

Boutaghou, Z. E., Erdman, A. G., Stolarski, H., " Dynamics of Flexible Beams and Plates in large overall Motions ", *Journal of Applied Mechanics,* Dec. 1992, vol. 59, pp. 991–999

Downer, J. D., Park, K. C., Chiou, J. C., " A Computational Procedure for Multibody Systems including Flexible Beam Modeling", *AIAA,* 90 – 1237, 1990, pp. 443 – 455

Kane, T. R., Ryan, R. R. and Banerjee, A. K., " Dynamics of a Cantilever Beam attached to a moving Base", *Journal of Guidance and Control,* 1987, vol. 10, pp. 139 – 151

Karkoub, Mansour, *"An Experimental Investigation of the Elasto-Dynamic Behavior of a 4-bar Mechanism"*, MS thesis, Department of Mechanical Engineering, University of Minnesota, 1990.

Liou, F. W., Erdman, A. G., " Experimental Motion Analysis of Flexible Mechanisms and Drive Systems using High speed Camera and Digital Imaging Techniques ", *Oklahoma State University, 10th Applied Mechanism Conference,* Dec 1987, vol III, session 8c – 4.

Malvern, L. E., *Introduction to the Mechanics of a Continuous Medium,* Prentice-Hall, Inc., Englewood Cliffs, N.J., 1969.

Simo, J. C., Vu – Quoc, L., " The Role of Nonlinear Theories in Transient Dynamic Analysis of Flexible Structures", *Journal of Sound and Vibration,* 1987, vol. 119, no. 3, pp. 487 – 509

Simo, J. C., " A Finite Strain Beam Formulation, The Three – Dimensional Dynamic Problem Part I", *Computer Methods in Applied Mechanics and Engineering,* 1985, vol. 49, pp. 55 – 70

Simo, J. C., Vu – Quoc, L., " On the Dynamics in Space of Rods undergoing Large Motions – A Geometrically Exact Approach ", *Computer Methods in Applied Mechanics and Engineering,* 1986, vol. 58, pp. 79 – 116

Stolarski, H., Belytschko, T., " Membrane Locking and Reduced Integration for Curved Elements ", *Journal of Applied Mechanics,* March 1982, vol. 49, pp. 172–176

Stolarski, H., Belytschko, T., " Shear and Membrane locking in Curved C^0 Elements ", *Computer Meth. Appl. Mech. Eng.,* vol. 41, pp. 279–296, 1983.

Timoshenko, S. and Goodier, N., *Theory of Elasticity,* Prentice-Hall, Inc., Englewood Cliffs, N.J., 1969.

Wang, Ming-Hsiung, *Application of the Mode Decomposition Technique in the Analysis of Curved Structures,* Ph.D. Thesis, University of Illinois at Chicago, 1989.

Wang, M., Stolarski, H., " A Mode Decomposition Technique for Nonlinear Analysis of Two-Dimensional Frames and Shells ", *in preparation.*(a)

Wang, M., Stolarski, H., " Large Deformation Analysis of Space Structures based on a Modified Strain Field ", *in preparation.*(b)

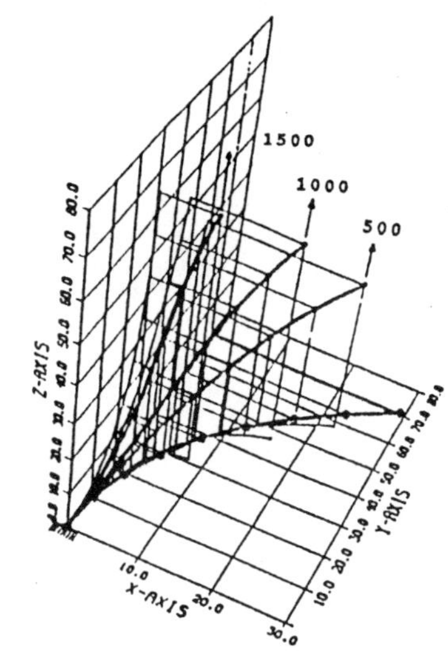

Figure 2: 45 deg Cantilever Bend

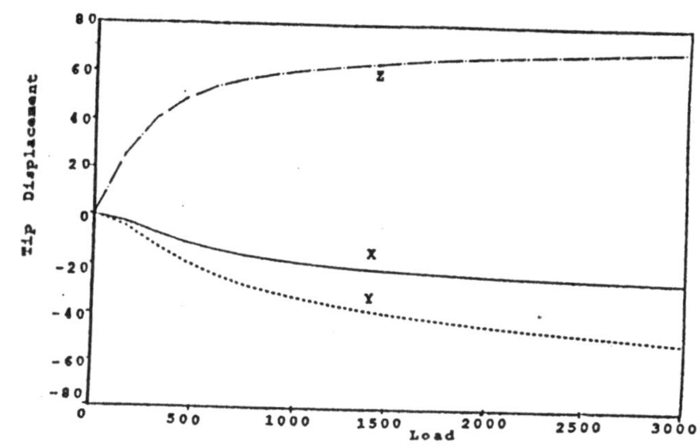

Figure 3: 45 deg Cantilever Bend–Load Displacement Curves

DE-Vol. 63, Vibrations of Mechanical Systems
and the History of Mechanical Design
ASME 1993

FREE VIBRATION OF RECTANGULAR MEMBRANES WITH LINEARLY VARYING TENSION AND LIGHT FLEXURAL RIGIDITY

D. J. Gorman
Department of Mechanical Engineering
University of Ottawa
Ottawa, Ontario, Canada

ABSTRACT

The proposal to utilize tensioned rectangular membranes as antennae in interspace communication has resulted in the focus of considerable attention on the free vibration of membranes with complicated in-plane loadings. In this paper the problem of analizing the vibration of a rectangular membrane with linear variation in tension in one direction is examined. This tension variation occurs due to gravitational forces when the membranes are tested experimentally in vibration laboratories. The Rayleigh-Ritz energy method is employed to obtain analytical results. Convergence is found to be rapid. Mode shapes are compared with those of known classical solutions when no tension variation is permitted. Generation of the eigenvalue matrix is demonstrated to be a very simple task.

INTRODUCTION

As a result of a proposal to utilize large tensioned rectangular membranes as antennae in inter-space communication, a strong interest has arisen in the free vibration frequencies and mode shapes of such structures. In their ultimate configuration these membranes will constitute a complicated structure. Typically, the aspect ratio will be about 2.5:1, the long edges being free while the short edges are tensioned by a system of discontinuous transverse bars. The bars, in turn, are anchored by short cables to fixed boundaries. A dense grid of small plate-like electronic cells will be attached to the membrane surface.

The membranes must be analyzed theoretically and experimentally to establish their free vibration characteristics before they can be accepted for deployment. While they will be tested experimentally under high vacuum they can only be tested while tensioned in a vertical plane. Two complications arise because of the existence of the attached electronic cells.

1) The average density of the membranes will be sufficiently high so that linear variations in tension due to gravitational forces cannot be neglected.

2) The same attached cells give the membrane an average flexural stiffness sufficiently high so that it cannot be neglected.

The stretched membrane then has light flexural rigidity and, in fact, could be thought of as a stretched, thin, highly flexible plate.

OBJECTIVE OF CURRENT STUDY

It was recognized that it would not be prudent to incorporate all of the complexities of the actual system into the initial studies. Accordingly, in the work reported here, an analysis is conducted for a membrane with linearly varying tension running in one direction and with non-negligible flexural rigidity. The boundary conditions are all of the classical simply supported edge conditions associated with plates. Uniform tension is applied to the short edges of the membrane only. It will be shown that a slight change in the functions utilized in the analysis permits examination of a membrane where the long edge boundary conditions much more closely approach the free edge condition.

MATHEMATICAL PROCEDURE

The membrane to be investigated here is shown schematically in Figure 1. It is tensioned in the vertical direction, the tension increasing linearly with y due to gravitational forces.

The tension is expressed mathematically as,

$$T(y) = T_0(1 + \alpha y/b) \qquad (1)$$

It will be appreciated that introducing this expression for tension into the governing differential equation results in the equation having one coefficient which is a function of the variable y. No analytical solution of the classical type is, therefore, achievable.

A solution is achieved, however, by means of the Rayleigh-Ritz method. Toward this

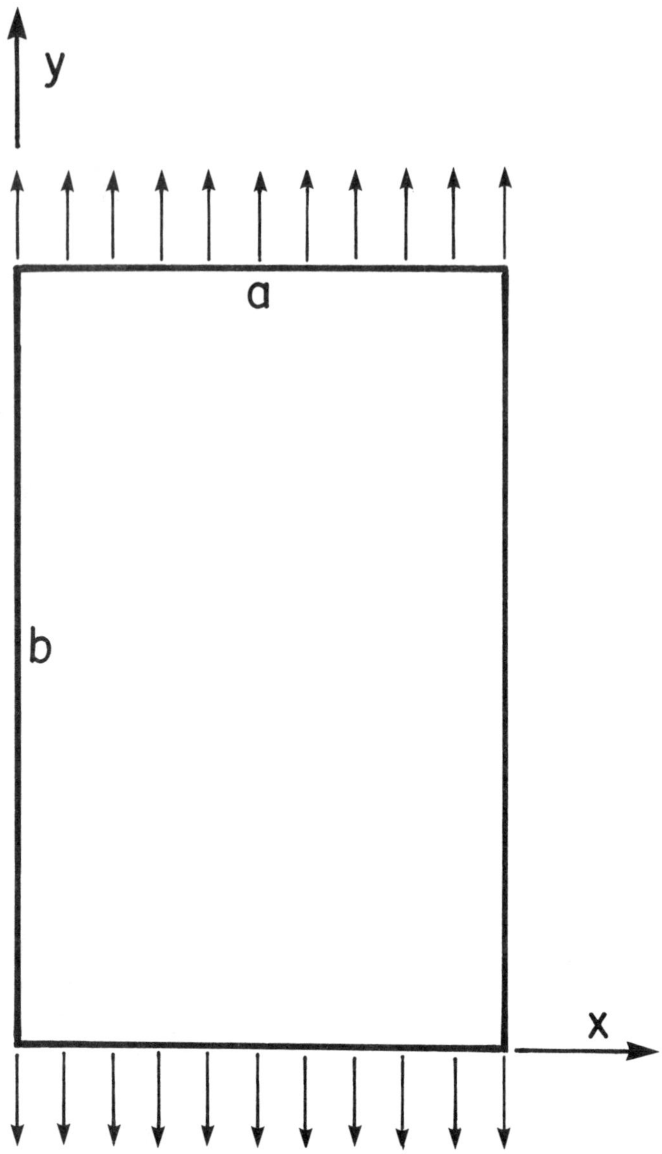

Figure 1. Schematic representation of rectangular membrane tensioned in vertical plane.

end the mode shape lateral displacement is taken as,

$$W(x,y) = \sum_{n=1}^{K} E_n \sin\frac{n\pi y}{b} \cdot \sin\frac{m\pi x}{a} \quad (2)$$

$$m=1,2,\ldots$$

It will be noted that all boundary conditions are satisfied and, with each value of m, m=1,2,..., there will be a family of mode shapes. In order to obtain these solutions we must obtain the total potential energy of the system and minimize it with respect to the coefficients E_n.

Beginning with the flexural energy of the system we have [1], (see list of symbols),

$$V_f = \frac{D}{2}\int_0^b\left(\int_0^a\left\{\left[\frac{\partial^2 W(x,y)}{\partial x^2} + \frac{\partial^2 W(x,y)}{\partial y^2}\right]^2 - 2(1-\nu)\right.\right.$$

$$\left.\left[\frac{\partial^2 W(x,y)}{\partial x^2}\frac{\partial^2 W(x,y)}{\partial y^2} - \left(\frac{\partial^2 W(x,y)}{\partial x \partial y}\right)^2\right]\right\} dx\,dy \quad (3)$$

Because of our choice of functions, it is shown that the second integral above vanishes [1].

Focusing on the first integral, it is easily shown that it may be written as,

$$V_F = \frac{D}{2}\int_0^b\left(\int_0^a \sin^2\frac{m\pi x}{a}\,dx\right)\sum_{n=1}^{K} E_n^2\left[\left(\frac{m\pi}{a}\right)^2 + \left(\frac{n\pi}{b}\right)^2\right]^2 \sin^2\frac{n\pi y}{b}\,dy \quad (4)$$

It will be noted that the cross products obtained on squaring the series are neglected here. Due to orthogonality of the functions selected, the integrals of these cross products would equal zero. Conducting the simple integration of equation (4) we obtain,

$$V_F = \frac{\pi^2 D}{8\phi b^2}\sum_{n=1}^{K} E_n^2[m^2\phi^2 + n^2]^2 \quad (5)$$

Turning next to the energy associated with tension, and referring to Timoshenko [1], we obtain,

$$V_T = \frac{T_0}{2}\int_0^b\int_0^a\left(1 + \frac{\alpha y}{b}\right)\left(\frac{\partial W(x,y)}{\partial y}\right)^2 dx\,dy \quad (6)$$

The first integral of Eqn (6) is easily obtained by substituting from Eqn (2) and performing the integration as was done in development of Eqn. (5).

We thereby obtain,

$$V_{T1} = \frac{T_0}{8\phi}\sum_{n=1}^{K} E_n^2(n\pi)^2 \quad (7)$$

Focusing on the second integral of Eqn. (6), it is found that we require,

$$V_{T2} = \frac{T_0\alpha}{2}\int_0^b\left(\int_0^a \sin^2\frac{m\pi x}{a}\,dx\right)\frac{y}{b}\left\{\sum_{n=1}^{K} E_n\frac{m\pi}{b}\right.$$

$$\left.\cos\frac{n\pi y}{b}\cdot\sum_{J=1}^{K} E_J\frac{J\pi}{b}\cos\frac{J\pi y}{b}\right\} dy \quad (8)$$

Here, because of the multiplier y/b, the cross products obtained on squaring the series cannot be neglected.

Conducting the integration with respect to x, Eqn. (8) can be written as,

$$V_{T2} = \frac{T_0 \alpha a}{4b} \sum_{n=1}^{K} \sum_{J=1}^{K} E_n E_J n\pi J\pi \int_0^b \frac{y}{b} \cos\frac{n\pi y}{b} \cos\frac{J\pi y}{b} d(\frac{y}{b}) \qquad (9)$$

Replacing the variable y/b with ξ, the integral above becomes,

$$XX = \int_0^1 \xi \cos n\pi\xi \ \cos J\pi\xi \, d\xi \qquad (10)$$

$$XX = \frac{1}{2}\left\{\frac{[\cos(n-J)\pi - 1]}{[(n-J)\pi]^2} + \frac{[\cos(n+J)\pi - 1]}{[(n+J)\pi]^2}\right. \qquad (11)$$

we may, therefore, write,

$$V_{T2} = \frac{T_0 \alpha}{4\phi} \sum_{n=1}^{K} \sum_{J=1}^{K} E_n E_J n\pi \, J\pi \, XX \qquad (12)$$

Finally, we turn to the contribution toward total potential energy associated with the inertia forces. This becomes,

$$V_I = -\rho\omega^2 \int_0^a \int_0^b W(x,y)^2 \, dx \, dy \qquad (13)$$

Substituting for W(x,y) and performing the simple integration, we obtain,

$$V_I = -\rho\omega^2 \frac{ab}{4} \sum_{n=1}^{K} E_n^2 \qquad (14)$$

GENERATION OF EIGENVALUE MATRIX

It is known that for dynamic equilibrium to exist the total potential energy must be a minimum with respect to variations of each coefficient E_n, i.e., for each n we may write,

$$\frac{\partial(T.P.E.)}{\partial E_n} = 0 \qquad (15)$$

This gives rise to K simultaneous homogenous algebraic equations relating the E_n's . The coefficient matrix of this set of equations constitutes our eigenvalue matrix. Contributions to total potential energy are represented by Eqns. (5),(7),(12) and (14).

It is found that all derivatives of these equations, except Eqn. (12), contribute to the diagonal elements of the matrix only. Concentrating on these diagonal terms we have, for each n,

$$A(n,n) = \frac{D}{4\phi b^2}[\phi^2(m\pi)^2 + (n\pi)^2]^2 + \frac{T_0}{4\phi}(n\pi)^2(1 + \alpha/2) - \rho\omega^2\frac{ab}{4} \qquad (16)$$

Here we have included the one term of Eqn (12), (n=J), which does contribute to the diagonal elements. With n=J, the integral XX introduced earlier equals 1/4. It should be noted that in Eqn. (14), only one

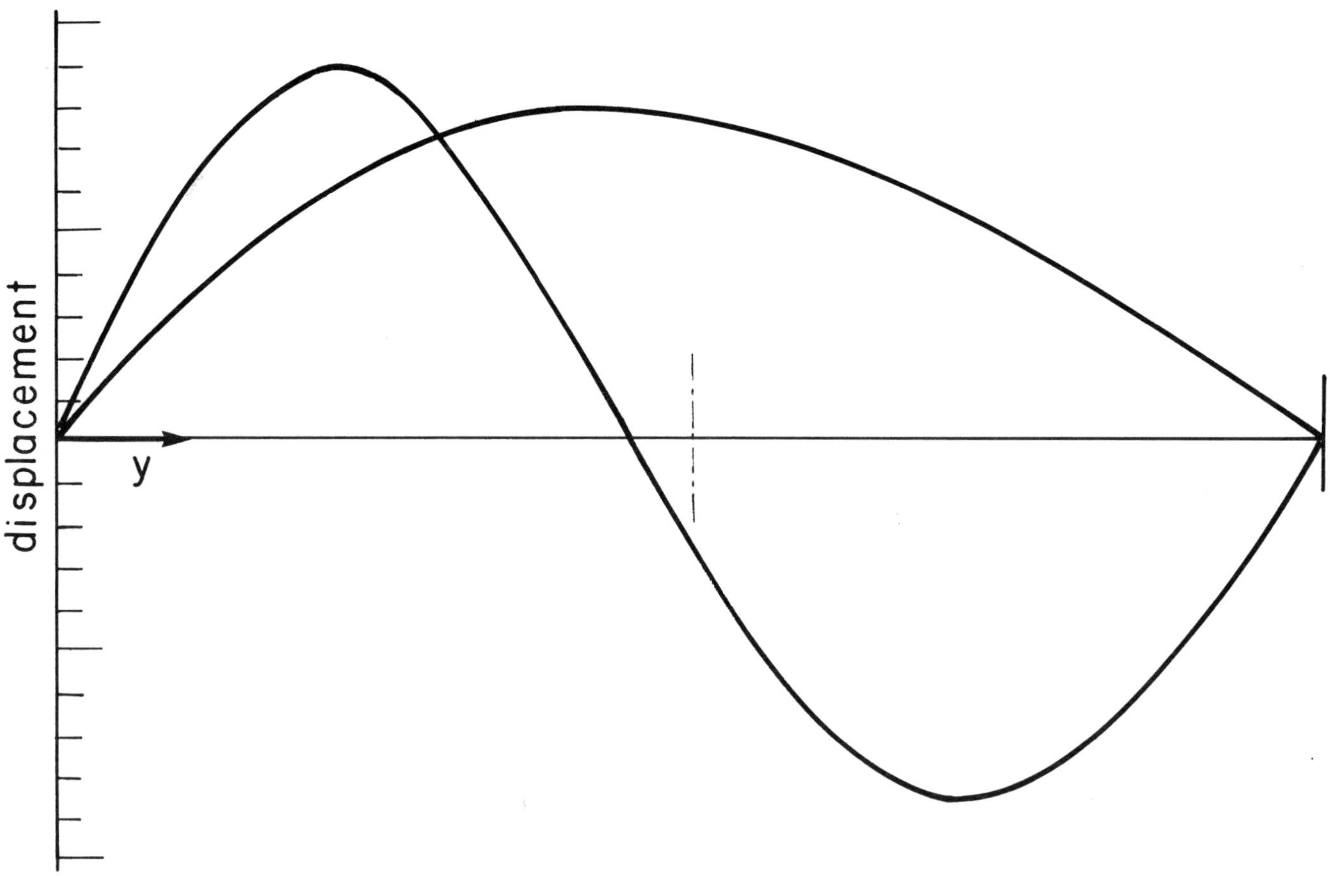

Figure 2. Plot of mode shape displacement along vertical central axis for first and second modes of tensioned membrane ($\phi=2.5$, m=1, $\alpha=12.0$, RAT=0.01).

of the E_n's is allowed to vary as the other E_n pertains to the inertia force which is not considered to vary during infinitesimal changes in the membrane configuration.

From Eqn. (12) we obtain (for $n \neq J$),

$$A(n,J) = A(J,n) = \frac{T_0 \alpha}{4\phi} n\pi J\pi XX \quad (17)$$

Eqns. (16) and (17) can be further simplified through multiplying each equation by $4\phi/T_0$ and introducing the eigenvalue $\lambda_T^2 = \omega b \sqrt{\rho/T_0}$, we then obtain, respectively,

$$A(n,n) = RAT[\phi^2(m\pi)^2 + (n\pi)^2]^2 + (n\pi)^2[1 + \alpha/2] - \lambda_T^4 \quad (18)$$

and

$$A(n,J) = A(J,n) = \alpha\, n\pi\, J\pi\, XX \quad (19)$$

where RAT, the stiffness ratio $=D/(b^2T_0)$.

Eqns. (18) and (19) permit generation of the eigenvalue matrix. The eigenvalues are obtained by searching for those values λ_T^2 which cause the determinant of the matrix to vanish. Mode shapes are obtained after setting one of the non-zero coefficients, E_n, equal to unity and solving for the remaining coefficients.

COMPUTED RESULTS

Before examining computed results, it is pointed out that certain limiting cases exist which allow verification tests to be performed on the above analysis. It will be appreciated that as the parameters α and T_0 are allowed to approach zero, the parameter λ_T^2 must approach the well-known eigenvalue for simply supported thin plates. Furthermore, as the stiffness ratio, as well as the parameter α, are allowed to approach zero, the eigenvalue must approach that of the classical membrane with unidirectional tention T_0. Both of these tests have been utilized to verify the analysis reported here.

In Figure 2 the results of a mode shape study are presented. The curves plotted in the figure represent modal displacement along the long central axis running in the y-direction. The parameter values are provided in the figure caption.

Concentrating on the first mode, it is observed that the deflection peaks at a point below the membrane horizontal centre line. Had there been no variation in tensile forces, we would have expected this deflection to take on the shape of a half sine wave.

It is noted that the displacement associated with the second mode also deviates from the full sine wave we would have expected had there been no gravitational forces. It crosses the axis a notable distance before the half-way mark along the membrane. We note also that the second spacial derivative of this latter curve takes on higher values in the lower section of the membrane. These higher derivatives permit increases in the elastic restoring forces which help compensate for reduction of restoring forces because of the lower tension forces in this region. In fact, the inertia forces, which must be balanced by the restoring forces, are slightly greater in the lower region because of the slightly higher amplitude of displacement.

CONCLUSIONS

Generation of the eigenvalue matrix for the problem under study here turns out to be a fairly simple operation. While the solution described does not resolve the more complex problem of ultimate interest, it constitutes a significant step in the right direction. It will be apparent to the reader that it would be a simple matter to replace the sine functions, sin $m\pi x/a$, of Equation (2) with trigonometric cosine functions. This would permit a very close simulation of free edge conditions at the extremities of these functions. In fact, the only deviation from free edge conditions would be related to a small residual bending moment along these edges because of the low flexural rigidity of the membrane. Satisfaction of free edge conditions will be investigated in future studies. The problem of incorporating the effects of rigid anchor bars into the analysis will also be studied.

NOMENCLATURE

a,b	membrane edge lengths
D	flexural rigidity
m	no. of half sine waves running across membrane
RAT	stiffness ratio = D/b^2T_0
T_o	tension at lower edge of membrane
T(y)	tension distributed throughout membrane
T.P.E.	total potential energy of system
V_F, V_T, V_I	contributions to membrane total potential energy associated with flexure, tension, and inertia forces, respectively
w	membrane lateral displacement
x,y	membrane spacial coordinates
α	parameter indicating rate of increase of tension with distance y
υ	Poisson's ratio
ϕ	membrane aspect ratio = b/a
ρ	mass of membrane per unit area
ω	circular frequency of membrane vibration
λ^2	eigenvalue based on membrane tension, equals $\omega b\sqrt{\rho/T_0}$

REFERENCE

1. Timoshenko, S. and Woinowsky-Krieger, S., "Theory of Plates and Shells", McGraw-Hill Book Company, 2nd Ed., (1959).

DE-Vol. 63, Vibrations of Mechanical Systems
and the History of Mechanical Design
ASME 1993

VIBRATION OF AN ECCENTRICALLY CLAMPED ANNULAR PLATE

Jung-Ge Tseng and Jonathan A. Wickert
Department of Mechanical Engineering
Carnegie Mellon University
Pittsburgh, Pennsylvania

ABSTRACT

Small amplitude vibration of an eccentric annular plate, which is free along its outer edge and clamped along the interior, is investigated through experimental and analytical methods. A disk with this geometry, or a stacked array in which the clamping and symmetry axes of each disk are nominally coincident, is common in data storage and brake systems applications. In the present case, the geometric imperfections on the boundary can have important implications for the disk's dynamic response. Changes that occur in the natural frequency spectrum, the mode shapes, and the free response under eccentric mounting are studied through laboratory measurements and an approximate discrete model of the plate. The natural frequencies and modes are found through global discretization of the Kamke quotient for a classical thin plate. For the axisymmetric geometry, the natural frequencies of the "sine" and "cosine" vibration modes for a specified number of nodal diameters are repeated. With increasing eccentricity, on the other hand, each pair of repeated frequencies splits at a rate that depends on the number of nodal diameters. Over a range of clamping and eccentricity ratios, the model's predictions are compared to the measured results.

INTRODUCTION

Annular plates are common structural elements, and particularly so in rotating machinery applications such as bladed disk assemblies, circular saw blades, brake systems, and disk drives for data storage. In principle, each disk is mounted such that the symmetry and rotation axes are nominally coincident, but a finite level of eccentricity is unavoidable, in part due to manufacturing and assembly tolerances. Further, in a disk drive that uses a removable storage medium, the disk's eccentricity cannot be controlled as carefully as that for a fixed media system. With removal and interchange, the disk becomes mounted on the spindle in an eccentric manner, so as to complicate tracking between the recording head and the disk. In addition to the "repeatable" component of the tracking error that is associated with run-out and the disk's rigid body rotation, the so-called "non-repeatable" component derives from the disk's flexible body vibration. The present study focuses on the latter, and in particular, on the variation of the plate's natural frequency spectrum and vibration modes under eccentric mounting.

A rich literature is available on the vibration of annular plates with coincident centers of the interior clamped edge and the exterior free edge [Vogel and Skinner, 1965; Leissa, 1969; Kim and Dickinson, 1989; and Weisensel, 1989]. In this axisymmetric case, a specified vibration mode (m, n) of the plate is classified according to the numbers m of nodal circles and n of nodal diameters. A pair of repeated natural frequencies exists for each $n \neq 0$. The corresponding modes are termed the "sine" and "cosine" components in recognition of the spatial distribution of the eigenfunctions in the circumferential direction; they are distinguished only by a phase shift. Each eigenfunction's radial component is a linear combination of regular and hyperbolic Bessel functions. On the other hand, when the plate is

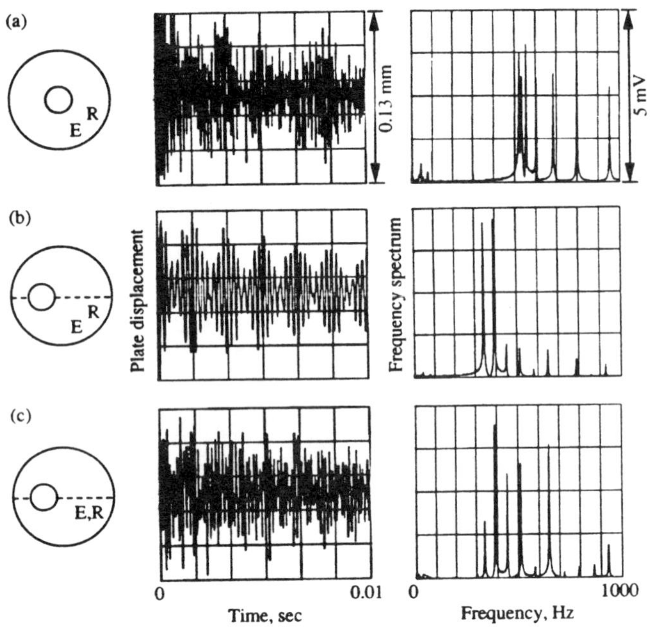

Figure 1: Measured responses and spectra for (a) an axisymmetric plate with non-collocated excitation E and response R points, (b) an eccentric plate with non-collocated E and R, and (c) an eccentric plate with collocated test points; $e^* = 15$ mm and $\alpha = 0.7$. Under eccentric mounting, the natural frequencies shift and split, and beating between adjacent vibration modes can occur.

mounted eccentrically so that the centers of the inner and outer edges are offset, the mode shapes and natural frequencies change in several respects.

Figure 1 shows measured responses of a test plate with an inner clamped edge of radius $a = 124$ mm and a free outer edge of radius $b = 178$ mm. The geometry and test procedure are further described below. The plate's responses are tabulated in the time and frequency domains for the three different configurations indicated by the sketches in the left column of Fig. 1. The plate was excited impulsively at point E, and the response was measured at R, for the axisymmetric configuration (case (a)) and under eccentric mounting with an offset of $e^* = 15$ mm (cases (b) and (c)). Two phenomenological observations can be made from this set of measurements. First, under some circumstances, beating in the transient time record can occur for the eccentric plate, but evidently beating does not occur in the axisymmetric case. Under impulsive loading, the plate in Fig. 1a responded with spectral content at many of its natural frequencies. However, the sine and cosine modes for a particular repeated frequency pair were not necessarily both excited. As a result, the response had a high modal content, and the time domain record has little discernible structure. Similar responses occur if E and R are collocated. For the eccentrically clamped plate, beating between two modes can occur, as indicated in Fig. 1b by the time record and by the spectral peaks at 341 Hz and 391 Hz which have nearly equal amplitude. Beating does not occur for some other combinations of E and R such as the collocated configuration depicted in Fig. 1c. As the second observation, the natural frequencies of the plate are lowered for this particular value of e^*, and evidently pairs of repeated frequencies split into distinct values.

Studies related to the vibration of annular plates with an eccentric inner boundary include [Nagaya, 1977a]. That solution required truncation of a Fourier series, and as a result the method was unable to predict higher vibration modes with fidelity. The companion vibration problem for membranes has also been studied [Nagaya, 1977b]. Liew [1992] used the Rayleigh-Ritz approach in Cartesian coordinates to obtain estimates of the lower frequencies for several annular plate configurations including eccentric line and ring supports.

In the present paper, the vibration of a thin annular disk under the geometric imperfection of an eccentric inner boundary is investigated experimentally and theoretically. For repeatability of the measurements, test plates were fixtured in a hydraulic press, and the displacements were measured with a non-contacting eddy current probe in order to eliminate frequency shifts which could be associated with the added mass of an alternative transducer. Changes in the natural frequency spectrum and vibration modes are evaluated over a range of the plate's clamping and eccentricity ratios. The plate is modeled through classical Kirchhoff theory, and an approximate discrete model which is amenable to numerical solution is obtained through global discretization of the Kamke quotient.

ECCENTRIC PLATE VIBRATION ANALYSIS

The governing equations for the plate model sketched in Fig. 2 are developed in dimensionless form. Two geometric parameters, the clamping ratio $\alpha = a/b$ and the eccentricity ratio $e = e^*/b$, define the plate's domain. Dissipation is not included in the model in light of the measured sharp spectral peaks which are present in Fig. 1. The center of the inner clamped edge is chosen as the origin of the polar coordinate system (r, θ) with basis vectors $\mathbf{e}_r$ and $\mathbf{e}_\theta$. The local normal $\mathbf{e}_n$ and tangent $\mathbf{e}_s$ vectors on the outer periphery are not aligned

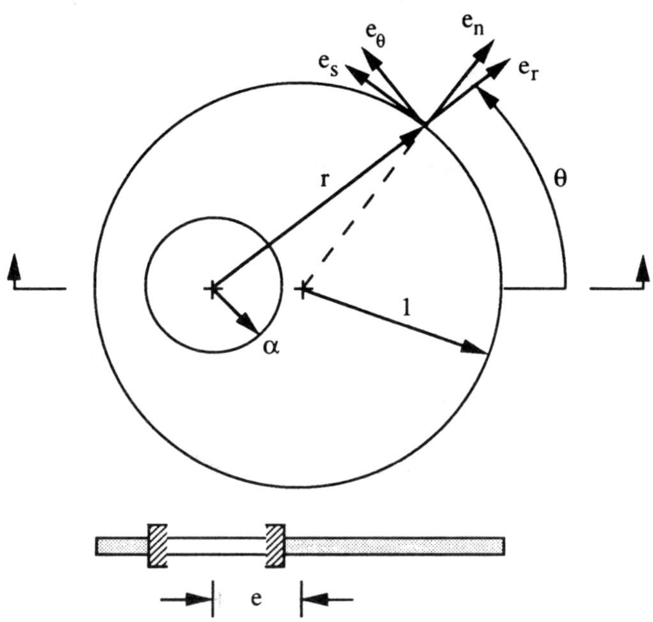

Figure 2: Schematic of the vibration model for an eccentrically clamped annular plate.

with the (r, θ) system, and the angular offset between these two coordinate systems is $\arcsin(e \sin\theta)$.

In accordance with standard thin plate theory, the eigenvalue problem for free linear vibration is

$$\nabla^4 W - \lambda^2 W = 0 \tag{1}$$

with solutions subject to the clamped conditions

$$W = 0 \quad , \quad \frac{\partial W}{\partial n} = 0 \tag{2}$$

at the inner edge $\{(r,\theta) : r = \alpha \text{ and } 0 \le \theta < 2\pi\}$, and to the free edge conditions

$$M_n = 0 \quad , \quad V_n = Q_n - \frac{\partial M_{ns}}{\partial s} = 0 \tag{3}$$

on $\{(r,\theta) : r = f(\theta) \text{ and } 0 \le \theta < 2\pi\}$. The function

$$f(\theta) = e\cos\theta + \sqrt{1 - e^2 \sin^2\theta} \tag{4}$$

defines the outer boundary. In dimensionless form, the eigenvalue in Eq. (1) is given by

$$\lambda^2 = \frac{12(1-\nu^2)\rho b^4}{Eh^3}(2\pi f)^2 \tag{5}$$

where ρ is the plate's mass per unit area, ν is Poisson's ratio, E is the elastic modulus, h is the plate's thickness, and f is the natural frequency. In Eqs. (2-3), M_n is the bending moment, M_{ns} is the twisting moment, and Q_n is the resultant shear resolved in the local normal-tangential system as explained in [Timoshenko and Woinowsky-Krieger, 1959]. In polar coordinates, the biharmonic operator is

$$\nabla^4 = \nabla^2\nabla^2 = \left(\frac{\partial^2}{\partial r^2} + \frac{1}{r}\frac{\partial}{\partial r} + \frac{1}{r^2}\frac{\partial^2}{\partial \theta^2}\right)^2. \tag{6}$$

Solutions of Eq. (1) are also extremals of the Kamke quotient

$$\mathcal{K}(W) = \frac{\mathcal{V}(W)}{\mathcal{T}(W)}, \tag{7}$$

which is interpreted as the ratio of the reference kinetic energy

$$\mathcal{T} = \frac{1}{2}\int_\Omega W^2 d\Omega \tag{8}$$

and potential energy

$$\mathcal{V} = \frac{1}{2}\int_\Omega (\nabla^2 W)^2 + 2(1-\nu)(\frac{1}{r}W_{,r\theta} - \frac{1}{r^2}W_{,\theta})^2 -2(1-\nu)W_{,rr}(\frac{1}{r}W_{,r} + \frac{1}{r^2}W_{,\theta\theta})d\Omega \tag{9}$$

of the plate for a specified spatial deformation W. Here the integration domain is defined by $\Omega = \{(r,\theta) : \alpha < r < f(\theta) \text{ and } 0 \le \theta < 2\pi\}$. Admissible W in Eq. (7) are elements of the function space that comprises all twice differentiable functions satisfying the essential conditions (2). $\mathcal{K}$ is derived from Rayleigh's quotient by successive application of Green's theorem in order to lower the differentiability requirements, and to relax the required boundary conditions, on W.

Approximate extremals of $\mathcal{K}$ are obtained through global Ritz discretization. The orientation of $\mathbf{e}_r$ and $\mathbf{e}_\theta$ chosen above simplifies the construction of an admissible expansion to the extent that the $r - \theta$ system is aligned with the $n - s$ system on the boundary where essential boundary conditions are specified. The displacement field is represented in terms of N generalized coordinates q_k and expansion functions $w_k(r,\theta)$

$$\begin{aligned} W(r,\theta) &= \sum_{k=1}^{N} q_k w_k \\ &= \sum_{i=1}^{NR}\sum_{j=1}^{N\theta} c_{ij} R_i(r)\Theta_j(\theta) \end{aligned} \tag{10}$$

with $N = (NR)(N\theta)$. Each w_k is separable in the r and θ coordinates, and is constructed from a product of terms

$$R_i = r^{i+1} \quad \text{for } i = 1, 2, 3, \cdots \tag{11}$$

$$\Theta_j = \begin{cases} \sin(j\theta) & \text{for } j = 1,3,5,\cdots \\ \cos(j\theta) & \text{for } j = 2,4,6,\cdots \end{cases} \tag{12}$$

as in the double summation of Eq. (10). In this manner, for a specified index k in the expansion term $q_k w_k$, the corresponding values of the indices i and j which are necessary for conversion to the separable representation $c_{ij} R_i \Theta_j$ become

$$i = k - \text{Int}\left(\frac{k-1}{N}\right) N$$
$$j = 1 + \text{Int}\left(\frac{k-1}{N}\right) N \tag{13}$$

Substitution of Eq. (10) into $\mathcal{K}$, and imposition of the stationarity requirement with respect to each generalized coordinate, lead to the discrete generalized eigenvalue problem $\underline{K}\underline{q} = \lambda^2 \underline{M}\underline{q}$. In particular, the (k, k') elements of the mass $\underline{M}$ and stiffness $\underline{K}$ matrices are given by

$$M_{kk'} = \int_0^{2\pi} \int_\alpha^{f(\theta)} r R_i R_{i'} \Theta_j \Theta_{j'} \, dr d\theta \tag{14}$$

$$\begin{aligned} K_{kk'} = \int_0^{2\pi} \int_\alpha^{f(\theta)} & r R_i'' R_{i'}'' \Theta_j \Theta_{j'} + \frac{1}{r} R_i' R_{i'}' \Theta_j \Theta_{j'} \\ & + \frac{1}{r^3} R_i R_{i'} \Theta_j'' \Theta_{j'}'' + \frac{2}{r^2} R_i' R_{i'} \Theta_j \Theta_{j'}'' \\ & + 2\ (1-\nu)\ (\frac{1}{r} R_i' R_{i'}' \Theta_j' \Theta_{j'}' + \frac{1}{r^3} R_i R_{i'} \Theta_j' \Theta_{j'}' \\ & - \frac{2}{r^2} R_i' R_{i'} \Theta_j' \Theta_{j'}') \\ & + 2\ \nu\ (R_i' R_{i'}'' \Theta_j \Theta_{j'} + \frac{1}{r} R_i'' R_{i'} \Theta_j \Theta_{j'}'')\, dr d\theta. \end{aligned} \tag{15}$$

Matrices $\underline{M}$ and $\underline{K}$ each have dimension $N \times N$. Corresponding elements of the sets $\{i, j, k\}$ and $\{i', j', k'\}$ in Eqs. (14-15) are related by Eqs. (13). Convergence of the method with increasing N was characterized, and the values $NR = 8$ and $N\theta = 15$ ($N = 120$) were used to obtain the results presented below. A 61 point Gauss-Kronrod scheme was used for quadrature in Eqs. (14-15), and an available QR algorithm provided the dimensionless frequencies λ. The mode shapes were reconstructed from the eigenvectors $\underline{q}$ and Eqs. (10-12).

In Fig. 3, the dependence of the plate's frequency spectrum on the eccentricity e is shown for the clamping ratio $\alpha = 0.7$. Only frequency loci that correspond to vibration modes with fewer than three nodal circles and five nodal diameters are shown for clarity; these loci are grouped according to common values of m. The natural frequencies for vibration modes with $n \neq 0$ split with

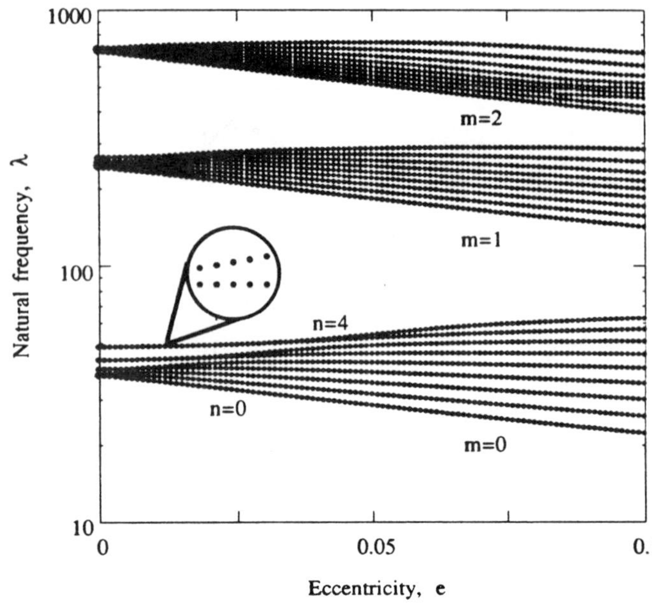

Figure 3: Dependence of the plate's predicted natural frequencies on the eccentricity; $\alpha = 0.7$, $m < 3$, and $n < 5$.

finite eccentricity, and the difference between these split values increases with e. Further, for a specified number of nodal circles, the rate at which the loci in a split frequency pair diverge near $e = 0$ decreases as number of nodal diameters grows. With regard to the $m = 0$ group, for instance, the fundamental mode at $e = 0$ is $(0, 0)$, followed by $(0, 1)$, $(0, 2)$, and so on. With increasing e, the frequency of the $(0, 0)$ mode decreases monotonically, as do the frequencies of the zero nodal diameter modes for larger values of m. The frequencies for all modes with $n \neq 0$, on the other hand, split into two distinct values from their repeated frequency values at $e = 0$. The spectral difference between two corresponding loci increases with e; for instance the separation between the two $(0, 1)$ frequencies at $e = 0.1$ is 17%. As indicated by the inset of Fig. 3 which details the split $(0, 4)$ pair near $e = 0.02$, vibration modes corresponding to larger values of n split at a slower rate than do the lower modes. The distinct frequency values for small e are still discernible in the vibration model, however.

The vibration modes for a pair of split frequencies are classified according to the dependence of the eigenfunctions on the θ coordinate. Modes (m, n) with $n \neq 0$ that are anti-symmetric about the line $\theta = 0$ of geometric symmetry are termed the split sine (S) modes, and

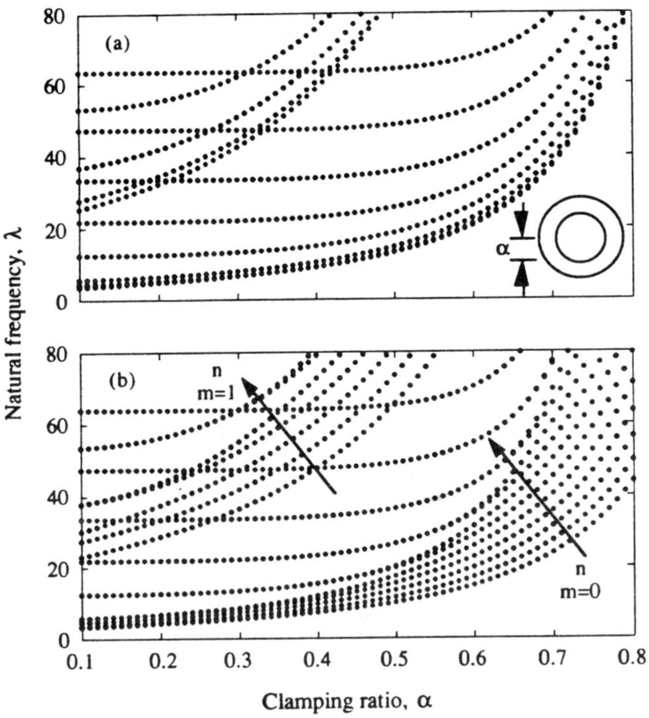

Figure 4: Dependence of the plate's predicted natural frequencies on the clamping ratio for (a) an axisymmetric plate, and (b) an eccentric plate with $e = 0.084$.

those which are symmetric are identified as the cosine (C) modes. In the general case, the spatial dependence of the split modes involves more than one harmonic in the circumferential direction, as in Eq. (10). However, in the limit of $e = 0$, the split modes asymptotically approach the repeated sine and cosine modes of the axisymmetric plate, which have only the nth harmonic present.

Variation of the natural frequency spectrum with respect to the clamping ratio α is illustrated in Fig. 4 for an axisymmetric plate, and for an eccentrically mounted one. Two groupings of loci that represent modes with zero and one nodal circles, with increasing nodal diameters n, are shown. In Fig. 4a, the $(0,0)$ and $(0,1)$ loci cross near $\alpha = 0.35$. In Fig. 4b, the difference between frequencies for the split vibration modes increases with α in a manner similar to the behavior with increasing eccentricity in Fig. 3. The rate at which a frequency pairs splits also decreases with α as the number of nodal diameters increases. In Fig. 5, results from the present model are compared for the case of $\alpha = 0.1$ with previously reported values. The two frequency estimates agree modestly for vibration modes up to $(0,2)$,

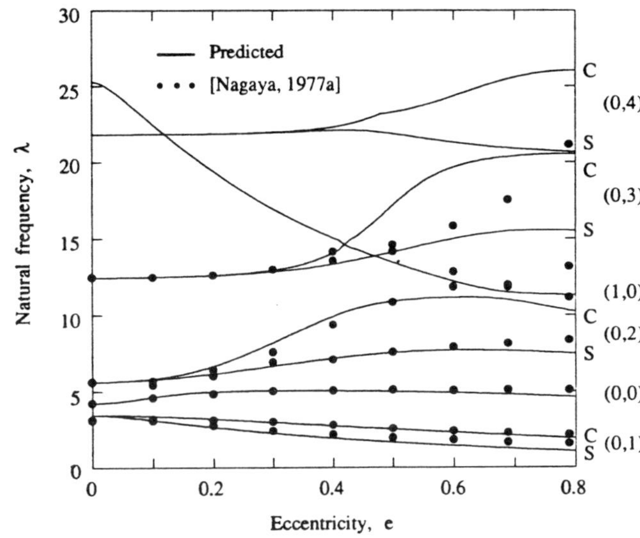

Figure 5: Predicted frequency spectrum for an eccentric annular plate; $\alpha = 0.1$. For large eccentricity, the $(1,0)$ frequency falls into the same range as the zero nodal circle modes with several nodal diameters. The "S" and "C" labels identify loci for the modes which are asymptotic to the repeated sine and cosine modes, respectively, at $e = 0$.

but there is some difference in the estimates of higher frequencies. In particular, for large e, the $(1,0)$ frequency falls into the same spectral range as the zero nodal circle modes with several nodal diameters. The (1,0) frequency locus, and its monotonically decreasing behavior, were not identified in [Nagaya, 1977a]. Further, in that paper, the frequency pair labeled "A3" and "S4," which correspond to the three nodal diameter modes, crossed such that the cosine component had a lower frequency than the sine component; this behavior is not borne out in the present measurements or analysis.

EXPERIMENT AND VERIFICATION

The experimental apparatus and instrumentation are sketched in Fig. 6. An annular aluminum plate with an inside radius of 84 mm, an outside radius of $b = 178$ mm, and a thickness of $h = 1.90$ mm was salvaged from a commercial computer hard disk drive. The plate was fixtured between two solid circular aluminum flanges with common outside radius $a = 124$ mm; this configuration provided a clamping ratio of $\alpha = 0.7$. The flanges were substantially thicker than the plate, and their

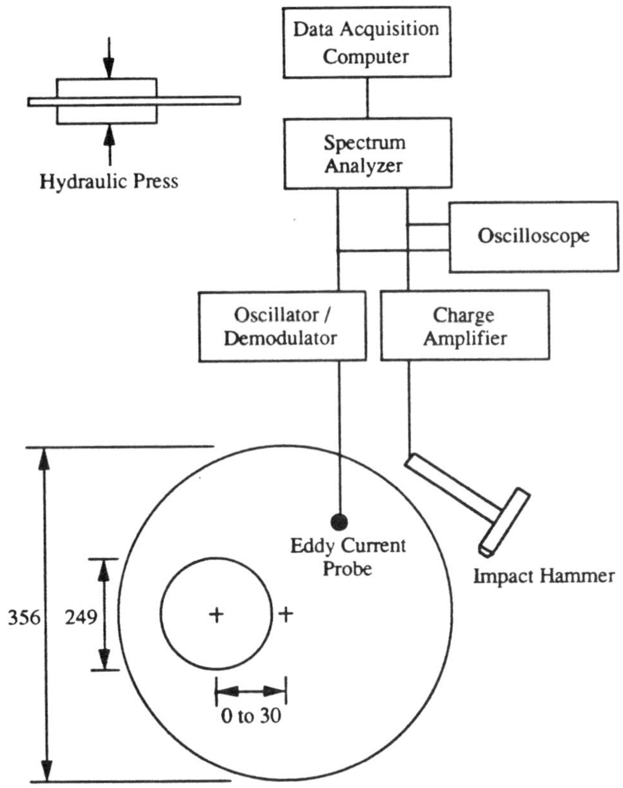

Figure 6: Schematic of the experimental apparatus (dimensions in millimeters). A hydraulic press pre-loads the plate and flange assembly to ensure uniformity of the contact pressure along the plate's inner boundary.

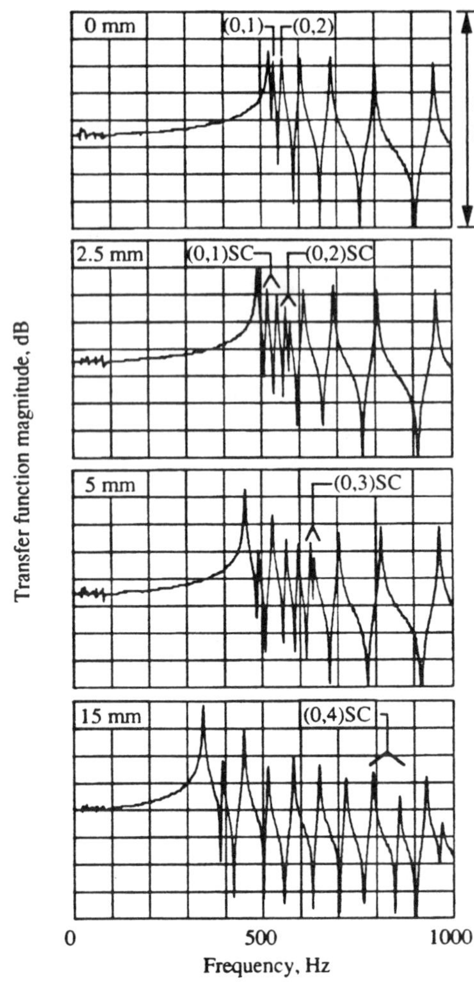

Figure 7: Measured transfer functions for an eccentrically clamped plate, detailing the progression of frequency splitting into sine S and cosine C values with increasing eccentricity; $e^* = 0$, 2.5, 5, and 15 mm. The excitation and response points are collocated.

elastic deformation relative to the plate was judged to be negligible during small amplitude vibration. The flanges were manually aligned and positioned on the plate at a specified value of e^* within the range 0 to 30 mm. The plate and flange assembly was pre-loaded in a hydraulic press to approximately 8 kN. The resulting distribution of contact pressure along the plate's inner boundary was sensibly uniform, and in the context of classical plate theory, it can be treated as a "clamped" boundary.

An eddy current probe with a linear range of approximately 1 mm was used to measure the displacement of the plate at a specified position (r, θ) without direct physical contact. In this manner, small changes in the plate's natural frequency spectrum and mode shapes were not introduced by, for instance, the added mass of an accelerometer. The plate was excited impulsively by a hammer instrumented with a piezoelectric force sensor which provided a linear signal up to 200 N. Displacement and force time records were recorded on an analog/digital storage oscilloscope, and a multichannel spectrum analyzer characterized the signals in the frequency domain. Peaks in the measured transfer functions provided the plate's natural frequencies. By transferring an ensemble of measurements to an available computer, the plate's vibration modes were determined through standard techniques with a commercial experimental modal analysis package.

Several transfer functions with collocated excitation and response measurement points are shown in Fig. 7. For the axisymmetric case ($e^* = 0$ mm), the $(0,0)$ mode has the lowest frequency, and each of the other spectral peaks shown within the 1 kHz frequency band repre-

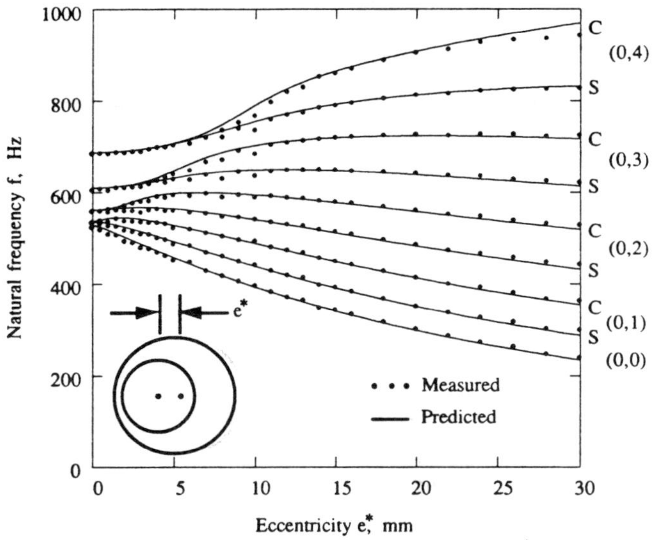

Figure 8: Comparison of measured and predicted natural frequencies for an eccentrically clamped plate; $\alpha = 0.7$.

sents repeated sine and cosine mode pairs up to $n = 6$. With $e^* = 2.5$ mm, for instance, splitting of both the $(0,1)$ and $(0,2)$ mode pairs is evident. In addition, the frequency of the $(0,0)$ mode has decreased to 494 Hz from its value of 528 Hz in the axisymmetric case. The vibration model predicts that all modes with $n \neq 0$ split for finite eccentricity, although for $e^* = 2.5$ mm, splitting of modes with $n > 2$ is not evident at the resolution chosen in Fig. 7. With $e^* = 5$ mm, on the other hand, the $(0,3)$ pair of split modes at 629 Hz (sine) and 639 Hz (cosine) can be seen. At $e^* = 15$ mm, which represents the dimensionless value $e = 0.084$, splitting of the $(0,4)$ and $(0,5)$ modes has occurred, and these split frequencies are no longer "closely" spaced. Measured and predicted natural frequencies of the test plate are compared in Fig. 8 for values of e^* up to 30 mm. In order to characterize the spectral behavior over a broad range of eccentricity, e^* was incremented in steps of 0.5 mm over $0 < e^* < 5$ mm, of 1 mm over $5 < e^* < 15$ mm, and of 2 mm for $e^* > 15$ mm. In particular, the coalescence of frequency loci in the limit of axisymmetric fixturing was investigated by choosing a small increment near $e^* = 0$. With $m = 0$ and $n < 5$, there are only five natural frequencies below 1kHz at $e^* = 0$, but there are nine well-spaced values when e^* reaches 30 mm. Except for the $(0,0)$ frequency which decreases monotonically with e^*, the other loci pass through a local maximum at a value that is greater than the corresponding frequency of the axisymmetric plate. For instance, the $(0,1)C$ frequency is maximized near $e^* = 1.5$ mm, and the $(0,2)C$ frequency is maximized near $e^* = 7$ mm.

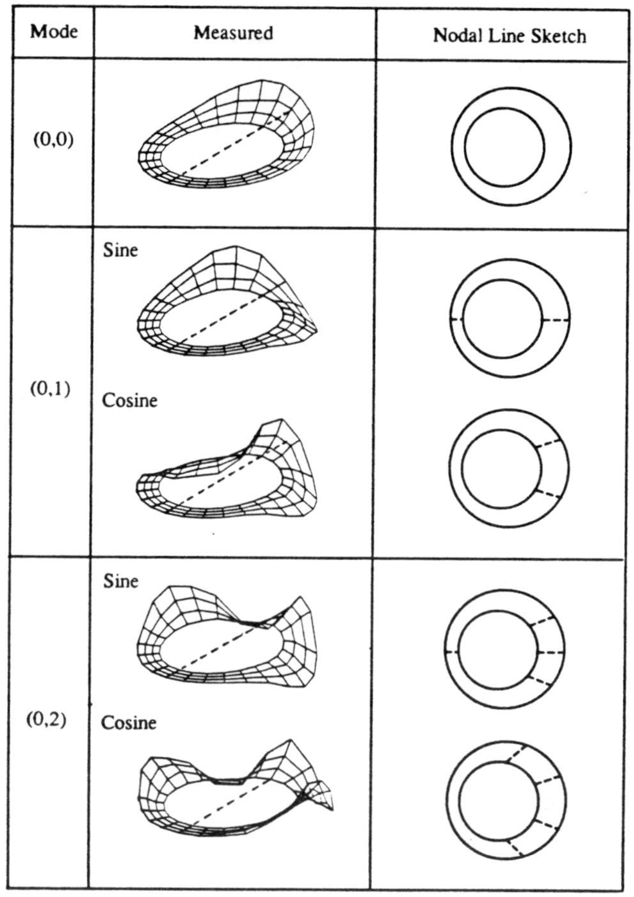

Figure 9: First five measured vibration modes and sketches of the nodal patterns; $e^* = 15$ mm ($e = 0.084$), and $\alpha = 0.7$.

The first five experimentally measured vibration modes, and sketches which indicate placement of the nodal lines, are shown in Fig. 9. Each sine mode retains as a nodal line the geometric symmetry axis of the plate's domain, namely the line that connects the centers of the inner clamped circular boundary and the free outer edge. For both the sine and cosine modes, the nodal lines become fixed in space by the asymmetry of the plate's domain. In the limiting case $e^* = 0$, the absolute orientation of the sine and cosine modes is arbitrary because of axisymmetry, and as a result, for an impulsive excitation, many modes of the plate will be excited, as indicated in Fig. 1a. With asymmetric fixturing, on the other hand, it is possible by judicious selection of E and R to excite one of the split modes, but not the other, because the nodal lines for the plate's sine and cosine modes become fixed. Further, when the split frequencies remain "closely" spaced for small e, beating between adjacent modes can occur. The time record in Fig. 1b is therefore associated with beating between the $(0, 0)$ and $(0, 1)S$ modes. This phenomenon is not associated with vibration of an axisymmetric plate.

SUMMARY

The vibration of an eccentrically clamped annular plate has been discussed from experimental and theoretical perspectives over a range of clamping and eccentricity ratios. The motivating applications and experimental configurations studied here are adequately described within the context of thin plate theory. The model that is obtained through global discretization of the Kamke quotient is useful to predict the plate's frequencies and vibration modes. Finite eccentricity which is present in clamping of the plate can affect its free and forced response, in part by increasing the number of resonant frequencies through splitting of repeated modes, and also by "locking" the orientation of the nodal lines with respect to the plate's axis of geometric symmetry. This distribution of frequencies and mode shapes can result in beating between closely spaced vibration modes.

ACKNOWLEDGMENT

This material is based (in part) upon work supported by the National Science Foundation under Grant Number MSS-9110638. The government has certain rights in this material.

REFERENCES

Kim, C.S., and Dickinson, S.M., 1989, "On the Lateral Vibration of Thin Annular and Circular Composite Plates Subject to Certain Complicating Effects", *Journal of Sound and Vibration*, Vol. 130, pp. 363-377.

Leissa, A.W., 1969, *Vibration of Plates*, NASA Report SP-160, pp. 7-35.

Liew, K.M., 1992, "Vibration of Eccentric Ring and Line Supported Circular Plates Carrying Concentrated Masses," *Journal of Sound and Vibration*, Vol. 156, pp. 99-107.

Nagaya, K., 1977a, "Transverse Vibration of a Plate Having an Eccentric Inner Boundary," *ASME Journal of Applied Mechanics*, Vol. 44, pp. 165-166.

Nagaya, K., 1977b, "Vibration of a Membrane Having a Circular Outer Boundary and an Eccentric Circular Inner Boundary," *Journal of Sound and Vibration*, Vol. 50, pp. 545-551.

Timoshenko, S.P., and Woinowsky-Krieger, S., 1959, *Theory of Plates and Shells*, McGraw-Hill.

Vogel, S.M., and Skinner, D.W., 1965, "Natural Frequencies of Transversely Vibrating Annular Plates," *ASME Journal of Applied Mechanics*, Vol. 32 , pp. 926-31.

Weisensel, G.N., 1989, "Natural Frequency Information for Circular and Annular Plates," *Journal of Sound and Vibration*, Vol. 133, pp. 129-134.

DE-Vol. 63, Vibrations of Mechanical Systems
and the History of Mechanical Design
ASME 1993

FORCED VIBRATION OF THICK VISCOELASTIC CYLINDERS

H. R. Hamidzadeh and G. R. Minor
Department of Mechanical Engineering
South Dakota State University
Brookings, South Dakota

ABSTRACT

Harmonic forced vibration of thick viscoelastic hollow cylinders of infinite extent is considered. The cylinder is excited by stresses applied at the inner and outer boundaries. The governing equation of motion is developed by utilizing three dimensional theory of elastodynamics. The material damping is allowed using complex elastic moduli for the viscoelastic medium. Modal displacements and stresses at any point in the medium are formulated in terms of boundary stresses. Frequency responses for radial, tangential and axial displacements are computed for different circumferential and axial wave numbers. The effect of different material loss factors on the frequency responses is examined for axial and non-axisymmetric modes. The dimensionless resonant frequencies for zero loss factor are compared with dimensionless natural frequencies available for elastic material. Comparison indicates excellent agreement between the results.

NOMENCLATURE

a,b = inner and outer radius
H = thickness of the cylinder
L = axial half wavelength
n = circumferential wave number
R = mean radius
r,θ,z = cylindrical coordinates
u = diplacement vector
V_1,V_2 = velocity of dilitation and distortional waves
ζ = wave number in axial direction
η = loss factor
μ,λ = Lame's elastic constants
ν = Poisson's ratio
ρ = density of the medium
$\underline{\sigma}$ = stress tensor
Ω = ω/ω_o
ω = exciting frequency
ω_o = $\pi V_2/L$

INTRODUCTION

In this paper we shall consider the formulation of harmonic forced vibrations of infinitely long viscoelastic thick cylinders. Considerable research has been performed for free vibrations of thick elastic cylinders, reported by Armenakas(1969) and Markus(1988). The work to be described here is to introduce a technique which has been found useful in the study of forced vibration of viscoelastic cylinders. Gazis(1959) and Armenakas et al(1969) have given solutions to the free vibration problem. In there formulation of the governing equations they were concerned with the transmission of elastic energy by means of elastic waves. Both references have introduced solutions by considering the eigen-value problem for traction free hollow surfaces. Armenakas et al extended his computational results and presented tables for dimensionless natural frequencies for different ratios of thickness to radius, and thickness to axial wave length for different circumferential wave numbers. Svardah(1984) treated wave propagation in a semi-infinite, elastic cylinders with traction free lateral surfaces initially at rest and excited by transient end loadings. Singal and Williams(1988) developed an approximate solutions for natural frequencies of thick cylinders and rings and verified their results by conducting experiments. Hamidzadeh and Chandler(1991) investigated circumferential harmonic vibrations of viscoelastically damped thick cylinders of infinite extent by employing two dimensional elastodynamic theory.

Stephen and Wang(1992) employed the Papleovitch-Neuber solution for the elastostatic diplacement equations of equilibrium and derived both axial and nonaxisymmetric solutions. They considered zero traction on the surface generators of he infinitely long elastic cylinder and formulated an eigen equation for free vibration. They computed smaller roots of the eigen equation for circumferential harmonic numbers of n = 0,1,2 and 3 for different wall thickness ratios.

The study reported here is directed at the evaluation of the harmonic response of thick viscoelastic cylinders. The excitations are six boundary stresses on inner and outer surfaces of the infinitely long cylinder. for the type of excitation described above, the responses will result from coupled and uncoupled radial, tangential and axial vibrations of the medium. The following analysis will show the procedure for determining the dimensionless resonant frequencies and the effect of material loss factor on the response for different circumferential wave numbers.

GOVERNING EQUATION

The displacements of any particle in a homogeneous, isotropic, viscoelastic medium using cylindrical coordinates (see Figure 1.) in the absence of body forces is given by equilibrium equations.

$$\mu\nabla^2\boldsymbol{u} + (\lambda+\mu)\nabla(\nabla\cdot \boldsymbol{u}) = \rho\frac{\partial^2\boldsymbol{u}}{\partial t^2} \tag{1}$$

The stress tensor $\underline{\sigma}$ is related to the displacement vector according to the following equation:

$$\underline{\sigma} = \lambda\nabla\cdot\boldsymbol{U}\boldsymbol{I}+\mu(\nabla\boldsymbol{U}+\boldsymbol{U}\nabla) \tag{2}$$

Where $\underline{I}$ is the unit tensor.

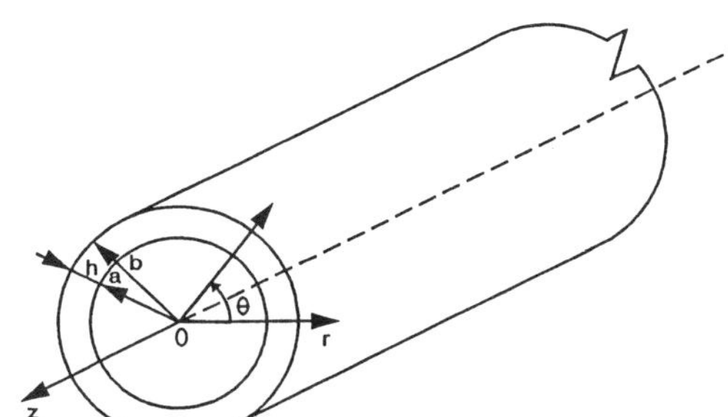

Figure 1. Reference coordinates and dimensions.

The equilibrium equation(1) can be solved by employing the potential functions ϕ and **H** such that

$$\boldsymbol{U} = \nabla\phi + \nabla\times\boldsymbol{H} \tag{3}$$

The potential functions are to satisfy the following wave equations:

$$V_1^2\ \nabla^2\phi = \frac{\partial^2\phi}{\partial t^2} \tag{4}$$

and

$$V_2^2\ \nabla^2\boldsymbol{H} = \frac{\partial^2\boldsymbol{H}}{\partial t^2} \tag{5}$$

The boundary conditions for the problem are stresses σ_{rr}, $\sigma_{r\theta}$ and σ_{rz} applied at both inner (r=a) and outer(r=b) boundaries. From the foregoing equations, solutions to the potential functions for harmonic frequency of ω and axial and circumferential wave numbers ζ and n are:

$$\phi = f(r)\cos n\theta \cos(\omega t+\zeta z) \quad (6\text{-a})$$

$$H_r = g_r(r)\, Sin\, n\theta \cos(\omega t+\zeta z) \quad (6\text{-b})$$

$$H_\theta = g_\theta(r)\cos n\theta \sin(\omega t+\zeta z) \quad (6\text{-c})$$

$$H_z = g_3(r)\sin n\theta \cos(\omega t+\zeta z) \quad (6\text{-d})$$

where,

$$f(r) = C_1 J_n(\alpha r) + C_2 Y_n(\alpha r) \quad (7\text{-a})$$

$$g_3(r) = C_3 J_n(\beta r) + C_4 Y_n(\beta r) \quad (7\text{-b})$$

$$\begin{aligned} 2g_1(r) &= g_r(r) - g_\theta(r) \\ &= 2C_5 J_{n+1}(\beta r) + 2C_6 Y_{n+1}(\beta r) \end{aligned} \quad (7\text{-c})$$

$$\begin{aligned} 2g_2(r) &= g_r(r) + g_\theta(r) \\ &= 2C_7 J_{n-1}(\beta r) + 2C_8 Y_{n-1}(\beta r) \end{aligned} \quad (7\text{-d})$$

and

$$\alpha^2 = \frac{\omega^2}{V_1^2} - \zeta^2 \quad (8\text{-a})$$

$$\beta^2 = \frac{\omega^2}{V_2^2} - \zeta^2 \quad (8\text{-b})$$

Using equations 1 and 2 the modal displacement and stress vectors are given by:

$$\{D\} = [CS][SN]\{C\} \quad (9)$$

where

$$\{D\} = \{u_r, u_\theta, u_z, \sigma_{rr}, \sigma_{r\theta}, \sigma_{rz}\}^T \quad (10)$$

$$\{C\} = \{C_1, C_2, C_3, C_4, C_5, C_6\}^T \quad (11)$$

Elements of 6×6 matrices [CS] and [SN] are given in the appendix. The modal boundary stresses can be presented by

$$\{BS\} = [CS][TN]\{C\} \quad (12)$$

where

$$TN_{i,j} = SN_{i+3,j} \quad (13\text{-a})$$

when r=a for i=1,3 and j=1,6

$$TN_{i+3,j} = SN_{i+3,j} \quad (13\text{-b})$$

when r=b for i=1,3 and j=1,6

Eliminating {C} between equations(9) and (12) yields

$$\{D\} = [SN][TN]^{-1}\{BS\} \quad (14)$$

For specific circumferential and axial wave numbers, equation(14) provides displacements, stresses and frequency responses for any particle in the medium in terms of amplitude of boundary stresses.

RESULTS AND DISCUSSION

A numerical solution has been obtained for forced vibrations of thick viscoelastic cylinders and dimensionless results are provided for different wave numbers. Figure 2. illustrates the variation of dimensionless resonant frequencies (resonant frequency factors Ω) for torsional and extensional modes (n=0) as well as nonaxisymmetric circumferetial modes (n=1,2,3 and 4) versus ratio of h/L. A thickness ratio of h/R=0.3, Poisson ratio of 0.3, and loss factor of η=0.0 are assumed. The results demonstrate the coupling effect between different thickness modes at particular values of h/L. To verify the solution some computed resonant frequency factors are compared with dimensionless natural frequencies provided by Armenakas et al(1969) in Table 1. The comparison establishes an excellent agreement between the results. Figures 3., and 4. illustrate the effect of material damping (loss factor) on the frequency responses for radial and tangential displacements for h/R = 0.3, Poisson's ratio of υ = 0.3 and h/L = 0.4. Presented responses are for material loss factors of η = 0.0, 0.1 and 0.5 for five consecutive circumferential wave numbers of 0, 1, 2, 3 and 4. The dimensionless amplitudes (amplitude factors) are defined as:

$$\begin{Bmatrix} A_r \\ A_\theta \\ A_z \end{Bmatrix} = \frac{\sigma}{aG} \begin{Bmatrix} U_r \\ U_\theta \\ U_z \end{Bmatrix} \qquad (15)$$

where U_i is the amplitude of u_i.

For n = 0 the peaks for radial and tangential displacements are occuring at different frequencies, this is because extensional and torsional modes are independent of one another. The nature of the non-axially symmetric modes can be determined by the diplacement responses at resonances. Results indicate that the addition of material damping to the system has been effective to attenuate the vibration amplitude.

CONCLUSION

An analytical procedure for evaluation of the harmonic forced vibrations of a thick visco-elastic cylinder is presented. Results are provided for response of both axial and non-axisymmetric modes of vibration. The effectiveness of material damping is studied and some results were satisfactorily compared with other availible results.

N = 0							
h/L		Resonant Frequency Factors					
0.0	Present	0.	0.1630	1.003	1.879	2.002	3.001
	Ref[1]	0.	0.1631	1.0034	1.8787	2.0018	3.0010
0.1	Present	0.1385	0.1895	1.021	1.857	2.031	3.001
	Ref[1]	0.1383	0.1895	1.0206	1.8568	2.0314	3.0008
0.5	Present	0.318	0.828	1.334	1.744	2.337	3.001
	Ref[1]	0.3179	0.8279	1.3344	1.7440	2.3367	3.0007
1.0	Present	0.749	1.420	1.912	2.003	2.829	3.079
	Ref[1]	0.7847	1.4200	1.9116	2.0029	2.8286	3.0788

N = 1							
h/L		Resonant Frequency Factors					
0.0	Present	0.	0.0955	0.225	1.0082	1.0343	1.858
	Ref[1]	0.	0.0958	0.2252	1.0082	1.0340	1.8581
0.1	Present	0.0613	0.165	0.260	1.0183	1.0454	1.842
	Ref[1]	0.0616	0.1648	0.2603	1.0183	1.0454	1.8423
0.5	Present	0.321	0.511	0.841	1.138	1.345	1.743
	Ref[1]	0.3210	0.5107	0.8414	1.1377	1.3452	1.7434
1.0	Present	0.804	1.005	1.4166	1.4362	1.9170	2.009
	Ref[1]	0.7886	1.0048	1.4167	1.4362	1.9168	2.0087

N = 2							
h/L		Resonant Frequency Factors					
0.0	Present	0.0361	0.192	0.351	1.0225	1.082	1.821
	Ref[1]	0.0360	0.1916	0.3515	1.0223	1.0821	1.8214
0.1	Present	0.055	0.228	0.382	1.0295	1.095	1.812
	Ref[1]	0.0546	0.2278	0.3825	1.0297	1.0949	1.8119
0.5	Present	0.333	0.540	0.880	1.151	1.377	1.743
	Ref[1]	0.3333	0.5401	0.8798	1.1506	1.3768	1.7426
1.0	Present	0.780	1.019	1.4176	1.4543	1.932	2.026
	Ref[1]	0.8002	1.0191	1.4177	1.4542	1.9322	2.0264

N = 3							
h/L		Resonant Frequency Factors					
0.0	Present	0.0955	0.287	0.493	1.045	1.155	1.787
	Ref[1]	0.0956	0.2872	0.4930	1.0445	1.1552	1.7865
0.1	Present	0.106	0.310	0.518	1.052	1.167	1.781
	Ref[1]	0.1064	0.3095	0.5177	1.0517	1.1674	1.7812
0.5	Present	0.359	0.584	0.938	1.172	1.427	1.744
	Ref[1]	0.3597	0.5837	0.9378	1.1722	1.4271	1.7438
1.0	Present	0.825	1.042	1.4255	1.4775	1.956	2.057
	Ref[1]	0.8196	1.0423	1.4254	1.4774	1.9565	2.0568

N = 4							
h/L		Resonant Frequency Factors					
0.0	Present	0.170	0.383	0.637	1.077	1.246	1.760
	Ref[1]	0.1700	0.3826	0.6370	1.0771	1.2464	1.7599
0.1	Present	0.179	0.398	0.659	1.083	1.257	1.757
	Ref[1]	0.1792	0.3983	0.6567	1.0828	1.2574	1.7572
0.5	Present	0.401	0.638	1.009	1.202	1.493	1.751
	Ref[1]	0.4015	0.6376	1.0089	1.2023	1.4926	1.7510
1.0	Present	0.845	1.073	1.439	1.505	1.988	2.100
	Ref[1]	0.8467	1.0735	1.4395	1.5049	1.9883	2.1005

Table 1. Comparison of dimensionless resonant and natural frequencies, h/R = 0.3.

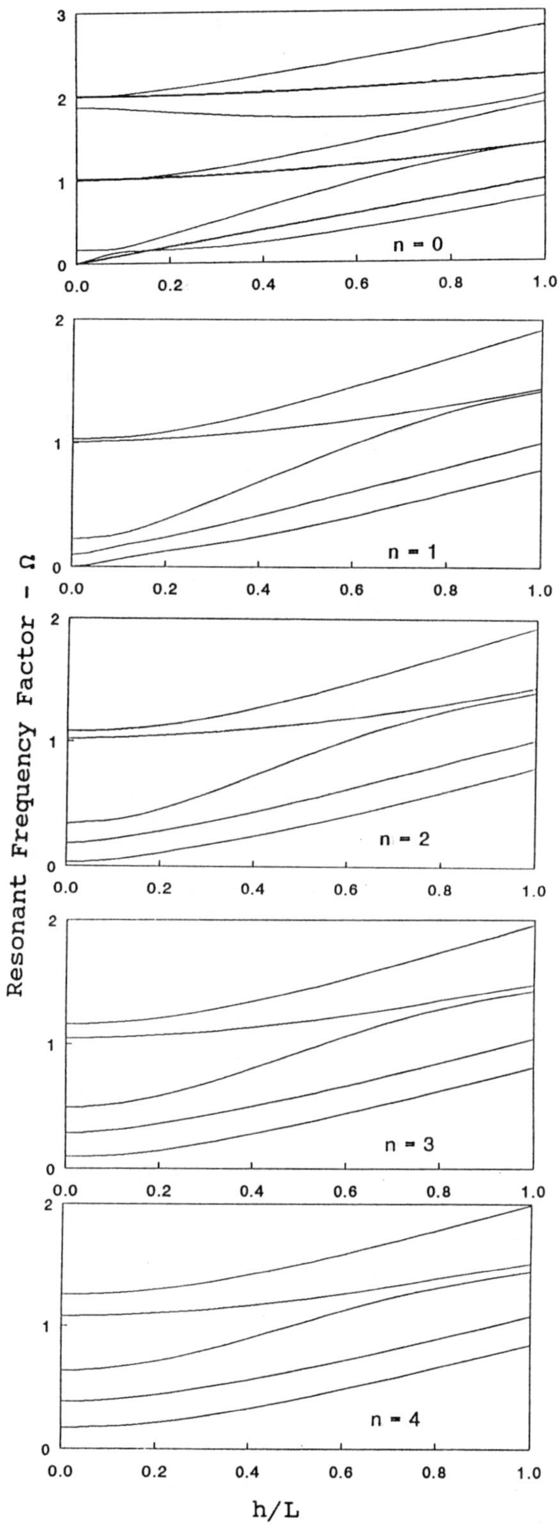

Figure 2. Variation of resonant frequency factors versus h/L for h/R=0.3.

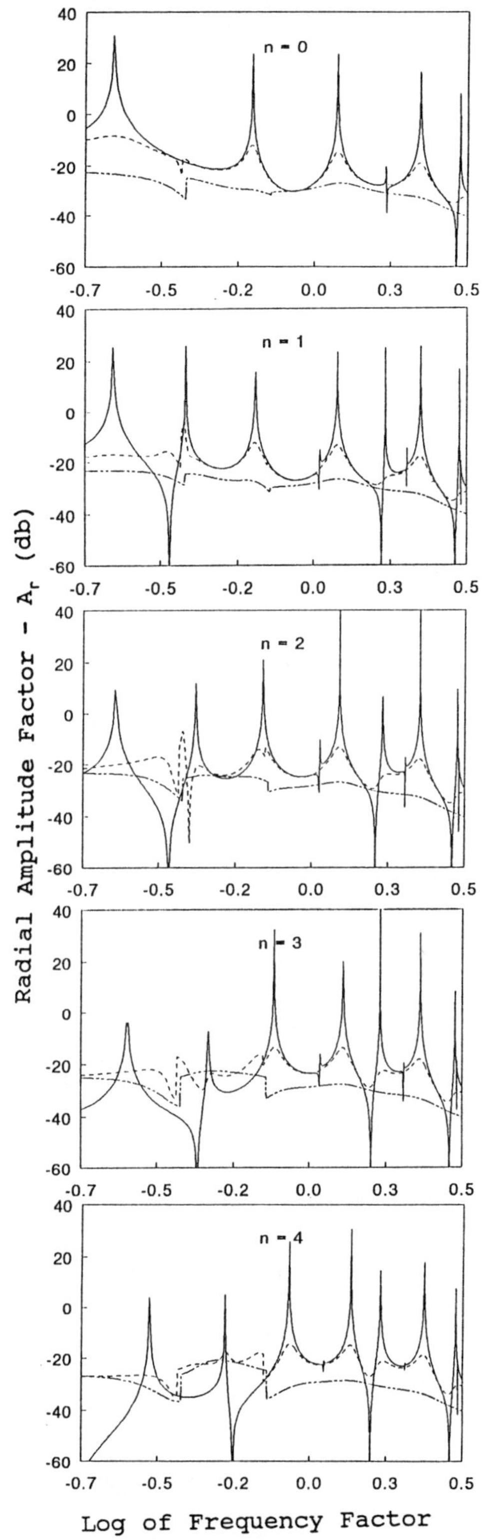

Figure 3. Frequency response for radial diplacement, h/R=0.3, h/L=0.4.

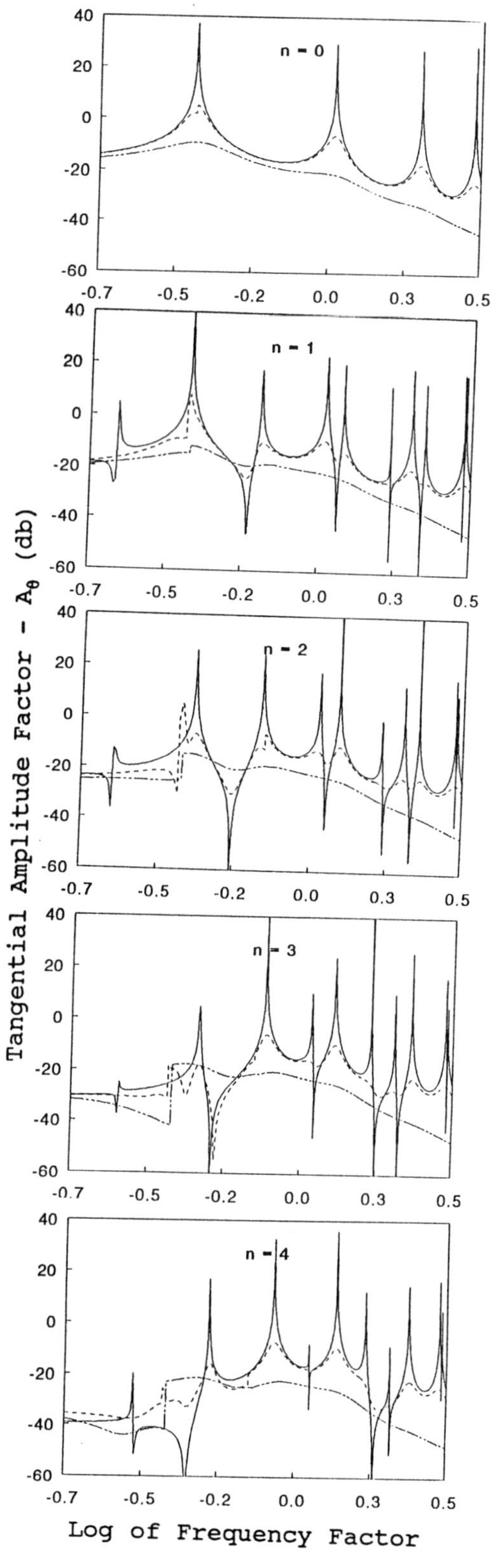

Figure 4. Frequency response for tangential displacement, h/R=0.3, h/L=0.4.

REFERENCES

Armenakas, A. E., Gazis, D. C. and Hermann, G., 1968, "Free Vibrations of Circular Cylindrical Shells", Oxford: Pergamon Press.

Gazis, D. C., 1959, "Three Dimensional Investigation of the Propagation of Waves of Hollow Circular Cylinders", Journal of the Acoustical Society of America, Vol. 31, pp. 568-578.

Singal, R. K. and Williams, K., 1988, "A Theoretical and Experimental Study of Vibrations of Thick Circular Cylindrical Shells and Rings", ASME Journal of Vibrations, Acoustics, Stress and Reliability in Design, Vol. 110, pp. 533-537.

Markus, S., 1988, "The Mechanics of Vibrations of Cylinrical Shells", Elsevier.

Hamidzadeh, H.R., and Chandler, D.E., 1991,"Circumferential Vibrations of Three Layered Sandwich Cylinders", The Proceedings of the Biennial ASME Conference on Mechanical Vibrations and Noise, Vol. DE-36, 233-237.

Stephen, N.G. and Wang, M.Z., 1992, "Decay Rates for the Hollow Circular Cylinder", Journal of Applied Mechanics, Vol. 59, 747-753.

Svardah, P.A., 1984, "Three-Dimensional Analysis of Axisymmetric Transient Waves in Hollow Elastic Cylinders", ASME Journal of Applied Mechanics, Vol. 51, pp. 792-796.

APPENDIX

The following are the elements of the matrices [CS] and [SN]. Only the nonzero elements are given.

$$CS_{ij} = 0, \quad when\ i \neq j$$

$$CS_{11} = CS_{44} = Cos\ n\theta\ Cos(\omega t + \zeta z)$$

$$CS_{22} = CS_{55} = \sin n\theta \cos(\omega t + \zeta z)$$

$$CS_{33} = CS_{66} = \cos n\theta \sin(\omega t + \zeta z)$$

$$SN_{11} = \alpha J_n/(\alpha r), \quad SN_{12} = \alpha Y_n/(\alpha r)$$

$$SN_{13} = \zeta J_{n+1}(\beta r), \quad SN_{14} = \zeta Y_{n+1}(\beta r)$$

$$SN_{15} = \frac{n}{r} J_n(\beta r), \quad SN_{16} = \frac{n}{r} Y_n(\beta r)$$

$$SN_{21} = -\frac{n}{r} J_n(\alpha r), \quad SN_{22} = -\frac{n}{r} Y_n(\alpha r)$$

$$SN_{23} = \zeta J_{n+1}(\beta r), \quad SN_{24} = \zeta Y_{n+1}(\beta r)$$

$$SN_{25} = -\beta J_n'(\beta r), \quad SN_{26} = -\beta Y_n'(\beta r)$$

$$SN_{31} = -\zeta J_n(\alpha r), \quad SN_{32} = -\zeta Y_n(\alpha r)$$

$$SN_{33} = -\beta J_{n+1}'(\beta r) - \frac{n+1}{r} J_{n+1}(\beta r)$$

$$SN_{34} = -\beta Y_{n+1}'(\beta r) - \frac{n+1}{r} Y_{n+1}(\beta r)$$

$$SN_{41} = -\lambda(\alpha^2+\zeta^2) J_n(\alpha r) + 2\mu\alpha^2 J_n''(\alpha r)$$

$$SN_{42} = -\lambda(\alpha^2+\zeta^2) Y_n(\alpha r) + 2\mu\alpha^2 Y_n''(\alpha r)$$

$$SN_{43} = 2\mu\zeta\beta J_{n+1}'(\beta r), \quad SN_{44} = 2\mu\zeta\beta Y_{n+1}'(\beta r)$$

$$SN_{45} = 2\mu\frac{n}{r}\beta J_n'(\beta r) - 2\mu\frac{n}{r^2} J_n(\beta r)$$

$$SN_{46} = 2\mu\frac{n}{r}\beta Y_n'(\beta r) - 2\mu\frac{n}{r^2} Y_n(\beta r)$$

$$SN_{51} = -2\mu\frac{n}{r}\alpha J_n'(\alpha r) + 2\mu\frac{n}{r^2} J_n(\alpha r)$$

$$SN_{52} = -2\mu\frac{n}{r}\alpha Y_n'(\alpha r) + 2\mu\frac{n}{r^2} Y_n(\alpha r)$$

$$SN_{53} = -\mu\zeta\frac{n+1}{r} J_{n+1}(\beta r) + \mu\zeta\beta J_{n+1}'(\beta r)$$

$$SN_{54} = -\mu\zeta\frac{n+1}{r} Y_{n+1}(\beta r) + \mu\zeta\beta Y_{n+1}'(\beta r)$$

$$SN_{55} = -\mu\frac{n^2}{r^2} J_n(\beta r) - \mu\beta^2 J_n''(\beta r) + \frac{\mu}{r}\beta J_n'(\beta r)$$

$$SN_{56} = -\mu\frac{n^2}{r^2} Y_n(\beta r) - \mu\beta^2 J_n''(\beta r) + \frac{\mu}{r}\beta Y_n'(\beta r)$$

$$SN_{61} = -2\mu\zeta\alpha J_n'(\alpha r), \quad SN_{62} = -2\mu\zeta\alpha Y_n'(\alpha r)$$

$$SN_{63} = \mu\left(\frac{n+1}{r^2} - \zeta^2\right) J_{n+1}(\beta r) - \mu\left(\frac{n+1}{r}\right)\beta J_{n+1}'(\beta r) - \mu\beta^2 J_{n+1}''(\beta r)$$

$$SN_{64} = \mu\left(\frac{n+1}{r^2} - \zeta^2\right) Y_{n+1}(\beta r) - \mu\left(\frac{n+1}{r}\right)\beta Y_{n+1}'(\beta r) - \mu\beta^2 Y_{n+1}''(\beta r)$$

$$SN_{65} = -\mu\zeta\frac{n}{r} J_n(\beta r), \quad SN_{66} = -\mu\zeta\frac{n}{r} Y_n(\beta r)$$

Where prime denotes differentiation with respect to the arguments of the Bessel function.

DE-Vol. 63, Vibrations of Mechanical Systems
and the History of Mechanical Design
ASME 1993

VIBRATION ANALYSIS OF A PARTIALLY COVERED DOUBLE SANDWICH CANTILEVER BEAM WITH CONCENTRATED MASS AT THE FREE END

C. Levy and Q. Chen
Department of Mechanical Engineering
Florida International University
Miami, Florida

ABSTRACT

The partially covered, sandwich-type cantilever with concentrated mass at the free end is studied. The equations of motion for the system modeled via Euler beam theory are derived and the resonant frequency and loss factor of the system are analyzed. The variations of resonance frequency and system loss factor for different geometrical and physical parameters are also discussed. Variation of these two parameters are found to strongly depend on the geometrical and physical properties of the constraining layers and the mass ratio.

NOMENCLATURE

A_i	cross-sectional area of layer, i
B_1	$\frac{1}{EA}(E_3A_3+E_4A_4)$
B_2	E_3A_3/EA
B_3	E_4A_4/EA
B_4	$\frac{1}{EA}[(E_1A_1+E_4A_4)h_{02} - E_1A_1h_{01})]$
B_5	$(E_1A_1h_{01} + E_3A_3h_{02})/EA$
B_6	$(E_1A_1 + E_3A_3)/EA$
B_7	$B_1h_{01} - B_2h_{02}$
B_8	$(E_1A_1 + E_4A_4)/EA$
D	$b[E_1h_1^3 + E_3h_3^3 + E_4h_4^3]/12$
D_u	$\frac{1}{12}E_1h_1^3$ bending rigidity of the original beam
EA	$E_1A_1+E_3A_3+E_4A_4$
E_i	Young modulus of the ith layer
G^*_{ci}	complex modulus of the ith layer, $G^*_{ci} = G_{ci}(1+j\eta_{ci})$
h_i	thickness of the ith layer
h_{01}	$h_{c2}+\frac{1}{2}(h_1+h_4)$
h_{02}	$h_{c1}+h_{c2}+h_1+\frac{1}{2}(h_3+h_4)$
j	$\sqrt{-1}$
k_n^*	complex characteristic value
L	length of the beam
L_1	length of the damping coverage
M_1,M_2	bending moment
m	mass density per unit length
P_i	axial force of the ith layer
S_1,S_2	shear force
t	time
u_i	axial displacement of the ith layer
w	transverse displacement
x	axial or longitudinal coordinate

Greek Symbols

γ_i	shear deformation of the ith layer
η	system loss factor
η_{ci}	core loss factor of the ith layer
ρ_i	mass density of the ith layer material
Ω	natural frequency of the system
Ω^*	complex natural frequency, $\Omega^* = \Omega(1+j\eta)^{1/2}$

INTRODUCTION

The vibration problem of a beam carrying a concentrated mass or concentrated masses and having arbitrary boundary conditions at the end supports is of great interest to the practical engineer. For example, this system may be considered as a model of a robot arm with a mass in its end effector. The vibrational aspects of this kind of system may be of great interest, when, due to some motion at the shoulder joint, the system is made to oscillate. Rao (1990) discussed the cantilever Euler beam with concentrated tip mass as one of the examples in his book. Pan (1965) studied the transverse vibration of an Euler beam carrying a system of heavy bodies. Rami Reddy and Amba-Rao (1973) solved the problem of vibrations of beams with non-classical boundary conditions (rotational and translational springs at the supports) using transfer matrices. Amba-Rao and Hussaini (1975) obtained a closed-form solution for the classical problem of a beam. The dynamic properties of structures carrying concentrated masses were studied by Laura (1977, 1987). Stokey and Zorowski (1959) solved the normal vibrations of a uniform plate carrying any number of finite masses. An approximate method to find the fundamental frequency of a restrained cantilever beam carrying a tip heavy body was presented by Gurgoze (1986). As a means of reducing this vibration, the beam may be covered by a viscoelastic material. The use of viscoelastic materials as means of increasing energy loss in beams and plates under flexural vibrations have been investigated by many authors. For example, Rao (1978) and Ditaranto (1965) obtained differential equations of motion for a sandwich beam. The effects of different boundary conditions were also studied in Rao (1978) and short beams were investigated by Rao (1977). Even though cantilever beams with tip masses have been scrutinized extensively, no papers have discussed the partially covered double sandwich cantilever with tip mass. Levy and Chen presented an analysis of the double, sandwich-type cantilever beam covered with a viscoelastic material. The model developed in that investigation will be employed, except that a mass at the end of the beam will be added.

In this paper, the partially covered double sandwich cantilever with concentrated mass at the free-end is studied. The equations of motion for the system are derived and the resonant frequency and system loss factor for different geometrical and physical parameters are also discussed.

EQUATIONS OF MOTION

The double sandwiched cantilever with concentrated mass at the free-end is shown in Fig.1. To make the mathematical method tractable, the following assumptions are made in the analysis:

1). The beam deflection is small and uniform across any section;

2). The primary beam and the upper and lower constraining beams are assumed to be isotropic;

3). The longitudinal and rotatory inertia effects at the beam are neglected;

4). The damping layers which carry shear but no direct stress are assumed to be linear viscoelastic;

5). No slip occurs at the interface between the layers;

6). The mass at the end effector can be modeled as a concentrated mass.

The beam is separated into two sections. Section 1 is a double sandwich-type beam containing the damping layers, and Section 2 is the ordinary beam with mass M_0 at its right-end. Matching conditions at the interface of the two sections will be applied on the transverse displacement and rotation, and force and moment continuity. The mass of the beam per unit length for Section 2 may be expressed as:

$$m_{20} = m_2 + M_0 \, \delta \, (x-L) \, , \qquad (1)$$

where m_2 is the mass per unit length of the uncovered portion of the beam, and δ (x-L) is the Dirac delta function. Hence, the differential equation of motion for Section 2 is:

$$D_{t1} \frac{\partial^4 w_2}{\partial x^4} + \left[m_2 + M_0 \delta (x-L) \right] \frac{\partial^2 w_2}{\partial t^2} = 0, \qquad (2)$$

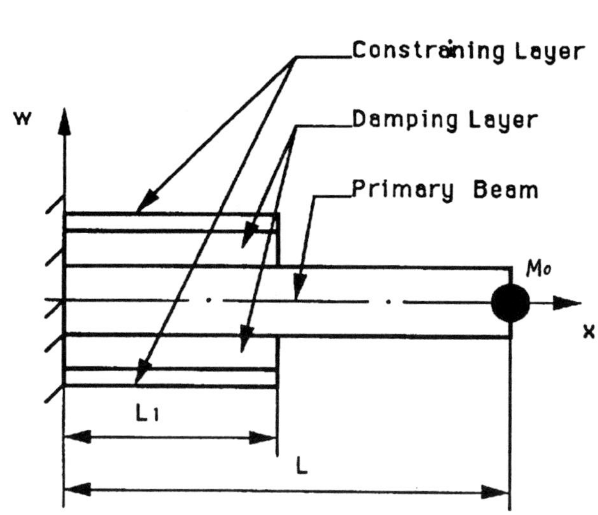

(a) THE COVERED BEAM WITH TIP MASS

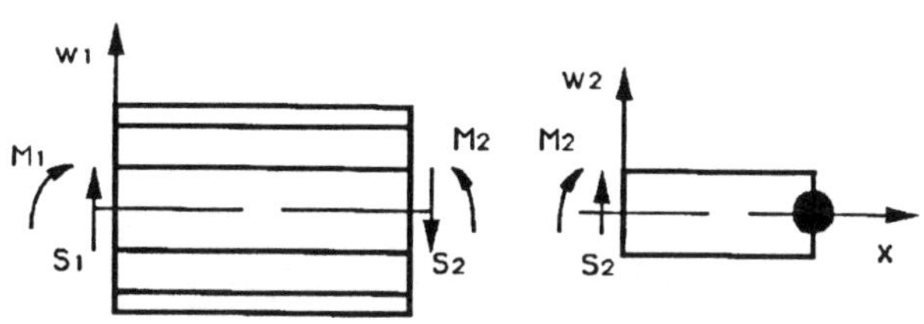

(b) THE TWO SECTIONS OF THE BEAM

FIG. 1

where $D_{t1}=E_1h_1^3b/12$ is the bending stiffness. For the interface between Sections 1 and 2, geometric continuity and generalized force continuity must apply. This leads to the following boundary conditions:

at $x=L_1$:

$$D_{t1}\frac{\partial^2 w_2}{\partial x^2} - M_2 = 0 \tag{3a}$$

$$D_{t1}\frac{\partial^3 w_2}{\partial x^3} + S_2 = 0 \quad . \tag{3b}$$

At x=L, since the effect of the mass is taken into account in Eq. (2), the boundary conditions are:

$$D_{t1}\frac{\partial^2 w_2}{\partial x^2} = 0 \tag{3c}$$

$$D_{t1}\frac{\partial^3 w_2}{\partial x^3} = 0 \quad . \tag{3d}$$

The differential equations of motion for Section 1 are given in Levy and Chen as:

$$\begin{bmatrix} a_{11} & a_{12} & a_{13} \\ a_{21} & a_{22} & a_{23} \\ a_{31} & a_{32} & a_{33} \end{bmatrix} \begin{bmatrix} w_1 \\ \gamma_{c1} \\ \gamma_{c2} \end{bmatrix} = \{0\} \ , \tag{4}$$

where the a_{ij}'s are the same as in Levy and Chen and are given in an appendix therein.

The corresponding boundary conditions are:

at x = 0:

$$(D + E_1A_1B_7^2 + E_3A_3B_4^2 + E_4A_4B_5^2)\frac{\partial^2 w_1}{\partial x^2}$$
$$+ (E_1A_1B_2B_7 - E_3A_3B_4B_8 + E_4A_4B_2B_5)h_{c1}\frac{\partial \gamma_{c1}}{\partial x}$$
$$+ (E_4A_4B_5B_6 - E_1A_1B_3B_7 - E_3A_3B_3B_4)h_{c2}\frac{\partial \gamma_{c2}}{\partial x}$$
$$+ M_1 = 0 \ \text{ or } \ \frac{\partial w_1}{\partial x} = 0 \tag{5}$$

$$(D + E_1A_1B_7^2 + E_3A_3B_4^2 + E_4A_4B_5^2)\frac{\partial^3 w_1}{\partial x^3}$$
$$+ (E_1A_1B_2B_7 - E_3A_3B_4B_8 + E_4A_4B_2B_5)h_{c1}\frac{\partial^2 \gamma_{c1}}{\partial x^2}$$
$$+ (E_4A_4B_5B_6 - E_1A_1B_3B_7 - E_3A_3B_3B_7)h_{c2}\frac{\partial^2 \gamma_{c2}}{\partial x^2}$$
$$- S_1 = 0 \ \text{ or } \ w_1 = 0 \tag{6}$$

$$(E_3A_3B_4B_8 - E_1A_1B_2B_7 - E_4A_4B_2B_5)h_{c1}\frac{\partial^2 w_1}{\partial x^2}$$
$$- (E_1A_1B_2^2 + E_3A_3B_8^2 + E_4A_4B_2^2)h_{c1}^2\frac{\partial \gamma_{c1}}{\partial x}$$
$$+ (E_1A_1B_2B_3 - E_3A_3B_3B_8 - E_4A_4B_2B_6)h_{c1}h_{c2}\frac{\partial \gamma_{c2}}{\partial x} = 0$$
$$\text{or } \ \gamma_{c1} = 0 \tag{7}$$

$$(E_1A_1B_3B_7 + E_3A_3\,B_3B_4 - E_4A_4B_5B_6)h_{c2}\frac{\partial^2 w_1}{\partial x^2}$$

$$+(E_1A_1B_2B_3 - E_3A_3B_3B_8 - E_4A_4B_2B_6)h_{c1}h_{c2}\frac{\partial \gamma_{c1}}{\partial x}$$

$$-(E_1A_1B_3^2 + E_3A_3B_3^2 + E_4A_4B_6^2)h_{c2}^2\frac{\partial \gamma_{c2}}{\partial x}=0$$

$$\text{or } \gamma_{c2} = 0\,. \tag{8}$$

Here, w_1 represents the displacement in Section 1, and the remaining symbols used in Eqs. (4-8) are defined in the Nomenclature. Letting M_1 and S_1 be replaced by $-M_2$ and $-S_2$, we obtain the boundary conditions for $x = L_1$.

RESONANT FREQUENCY AND LOSS FACTOR

As part of the design process of a partially covered, double sandwich-type cantilever beam, one will normally be interested in the resonant frequencies and loss factors in the first few modes of vibrations Rao (1978). To obtain the resonant frequencies and loss factors, assume the solution of Section 2 to be in the form:

$$w_2\,(x,t) = W_2(x)e^{j\Omega^* t} \tag{9}$$

For mass M_0 attached at $x = \xi$, substituting Eq. (9) into Eq. (2) gives:

$$\frac{d^4W_2(x)}{dx^4} - \frac{\omega^4}{L_2^4}\left[1+\frac{M_0}{m_0}\,L_2\delta(x-\xi)\right]W_2(x)=0, \tag{10}$$

where $m_0 = m_2L_2$, $L_2 = L-L_1$, $c_0^4 = D_{t1}/m_2$, and $\omega^2=\Omega^* L_2^2/c_0^2$. Taking the Laplace transform of Eq. (10) with respect to the variable, x yields:

$$\overline{W}_2(s)=[s^3W_2(L_1)+s^2W_2'(L_1)+sW_2''(L_1)+W_2'''(L_1)$$

$$+\frac{\omega^4}{L_2^3}\,\frac{M_0}{m_0}\,e^{-\xi s}\,W_2(\xi)]\bigg/\left[s^4-\frac{\omega^4}{L_2^4}\right], \tag{11}$$

where s represents the transform variable, and the first four terms give the boundary condition effects generated by the Laplace transform. Taking the inverse transform of Eq. (11), we obtain:

$$\begin{aligned}W_2(x) = W_2(\xi)\Bigg\{&\hat{W}_2(L_1)U\left[\frac{\omega x}{L_2}\right] + \hat{W}_2'(L_1)\left[\frac{L_2}{\omega}\right]V\left[\frac{\omega x}{L_2}\right] + \hat{W}_2''(L_1)\left[\frac{L_2}{\omega}\right]^2 S\left[\frac{\omega x}{L_2}\right]\\ &+ \hat{W}_2'''(L_1)\left[\frac{L_2}{\omega}\right]^3 T\left[\frac{\omega x}{L_2}\right] + \frac{M_0}{m_0}\,\omega T\left[\frac{\omega(x-\xi)}{L_2}\right]H(x-\xi)\Bigg\},\end{aligned} \tag{12}$$

where

$$\hat{W}_2^{(j)}(L_1) = W_2^{(j)}(L_1)/W_2(\xi)\ ,$$

$$U(x) = \frac{1}{2}\,[\cos(x)+\cosh(x)]\ ,$$

$$V(x) = \frac{1}{2}\,[\sin(x)+\sinh(x)]\ ,$$

$$S(x) = \frac{1}{2}\,[\cosh(x)-\cos(x)]\ ,$$

$$T(x) = \frac{1}{2}\,[\sinh(x)-\sin(x)]\ ,$$

and H(x) is the Heaviside step function. Since the solution must hold at $x = \xi$, Eq. (12) yields the following form of the frequency equation:

$$\begin{aligned}&\hat{W}_2(L_1)U\left[\frac{\omega_n\xi}{L_2}\right] + \hat{W}_2'(L_1)\left[\frac{L_2}{\omega_n}\right]V\left[\frac{\omega_n\xi}{L_2}\right] + \hat{W}_2''(L_1)\left[\frac{L_2}{\omega_n}\right]^2 S\left[\frac{\omega_n\xi}{L_2}\right]\\ &+ \hat{W}_2'''(L_1)\left[\frac{L_2}{\omega_n}\right]^3 T\left[\frac{\omega_n\xi}{L_2}\right] = 1\,.\end{aligned} \tag{13}$$

The corresponding eigenfunction is:

$$W_{2n}(x) = \hat{W}_{2n}(L_1)\, U\left(\frac{\omega_n x}{L_2}\right)$$

$$+ \hat{W}_{2n}'(L_1)\left(\frac{L_2}{\omega_n}\right) V\left(\frac{\omega_n x}{L_2}\right)$$

$$+ \hat{W}_{2n}''(L_1)\left(\frac{L_2}{\omega_n}\right)^2 S\left(\frac{\omega_n x}{L_2}\right) \qquad (14)$$

$$\hat{W}_{2n}'''(L_1)\left(\frac{L_2}{\omega_n}\right)^3 T\left(\frac{\omega_n x}{L_2}\right)$$

$$+ \frac{M_0}{m_0}\,\omega_n T\left[\frac{\omega_n(x-\xi)}{L_2}\right] H(x-\xi)\ ,$$

where $\hat{W}_{2n}(L_1)$, $\hat{W}_{2n}'(L_1)$, $\hat{W}_{2n}''(L_1)$, and $\hat{W}_{2n}'''(L_1)$ will be determined from the boundary conditions. According to the results found in Levy and Chen, the characteristic equation for section 1 is expressed as:

$$\det\begin{bmatrix} \bar{a}_{11} & \bar{a}_{12} & \bar{a}_{13} \\ \bar{a}_{21} & \bar{a}_{22} & \bar{a}_{23} \\ \bar{a}_{31} & \bar{a}_{32} & \bar{a}_{33} \end{bmatrix} = 0 \qquad (15)$$

where

$$\bar{a}_{11} = -\ m\Omega_n^{*2} + (D + E_1A_1B_7^2 + E_3A_3B_4^2 + E_4A_4B_5^2)k_n^{*4}$$

$$\bar{a}_{12} = +\ (E_1A_1B_2B_7 - E_3A_3B_4B_8 + E_4A_4B_2B_5)h_{c1}\ k_n^{*3}$$

$$\bar{a}_{13} = -\ (E_1A_1B_3B_7 - E_3A_3B_3B_4 + E_4A_4B_5B_6)h_{c2}\ k_n^{*3}$$

$$\bar{a}_{21} = (E_1A_1B_2B_7 - E_3A_3B_4B_8 + E_4A_4B_2B_5)h_{c1}\ k_n^{*3}$$

$$\bar{a}_{22} = (B_2^2E_1A_1 + B_8^2E_3A_3 + B_2^2E_4A_4)h_{c1}^2\ k_n^{*2} - G_{c1}A_{c1}$$

$$\bar{a}_{23} = (-\ E_1A_1B_2B_3 + E_3A_3B_3B_8 + E_4A_4B_2B_6)h_{c1}h_{c2}\ k_n^{*2}$$

$$\bar{a}_{31} = (E_4A_4B_5B_6 - E_1A_1B_3B_7 - E_3A_3B_3B_4)h_{c2}\ k_n^{*3}$$

$$\bar{a}_{32} = (E_3A_3B_3B_8 - E_1A_1B_2B_3 + E_4A_4B_2B_6)h_{c1}\ h_{c2}\ k_n^{*2}$$

$$\bar{a}_{33} = (E_1A_1B_3^2 + E_3A_3B_3^2 + E_4A_4B_6^2)h_{c2}^2\ k_n^{*2} - G_{c2}\ A_{c2}\ .$$

The reader is directed to the Nomenclature for the definitions of the symbols used in Eq. (15). In the same manner as in Levy and Chen, the various boundary conditions and continuity conditions may be expressed:

at x = 0:

$$\sum_{i=1}^{8} \bar{A}_{ni}\, k_{ni}^{*} = 0 \qquad (16)$$

$$\sum_{i=1}^{8} \bar{A}_{ni} = 0 \qquad (17)$$

$$\sum_{i=1}^{8} f_{ni}\, \bar{A}_{ni} = 0 \qquad (18)$$

$$\sum_{i=1}^{8} g_{ni}\, \bar{A}_{ni} = 0 \qquad (19)$$

The definitions of the $\bar{A}_{ni}$'s, f_{ni} and g_{ni} are taken from [12] and reproduced in Appendix I for completeness.

at x = L_1:

The moment balance at the interface, as viewed from Section 1, leads to the following:

$$\sum_{i=1}^{8} (D + E_1A_1B_7^2 + E_3A_3B_4^2 + E_4A_4B_5^2)\bar{A}_{ni}k_{ni}^{*2}\ e^{k_{ni}^{*}L_1}$$

$$+ \sum_{i=1}^{8} (E_1A_1B_2B_7 - E_3A_3B_4B_8$$

$$+ E_4A_4B_2B_5)h_{c1}f_{ni}\bar{A}_{ni}k_{ni}^{*}\ e^{k_{ni}^{*}L_1}$$

$$+ \sum_{i=1}^{8} (E_4A_4B_5B_6 - E_1A_1B_3B_7$$

$$-\ E_3A_3B_3B_4)h_{c2}g_{ni}\bar{A}_{ni}k_{ni}^{*}\ e^{k_{ni}^{*}L_1}$$

$$+ M_2 = 0 \qquad (20)$$

The moment balance at the interface, as viewed from Section 2, leads to:

$$
D_{t1}\left[\hat{W}_{2n}(L_1)\left(\frac{\omega_n}{L_2}\right)^2 S\left(\frac{\omega_n L_1}{L_2}\right) + \hat{W}_{2n}'(L_1)\left(\frac{\omega_n}{L_2}\right) T\left(\frac{\omega_n L_2}{L_2}\right) + \hat{W}_{2n}''(L_1)\, U\left(\frac{\omega_n L_1}{L_2}\right) + \hat{W}_{2n}'''(L_1)\left(\frac{L_2}{\omega_n}\right) V\left(\frac{\omega_n L_1}{L_2}\right)\right] - M_2 = 0 \tag{21}
$$

The shear force balance at the interface, as viewed from Section 1, leads to the following:

$$
\sum_{i=1}^{8} (D+E_1A_1B_7^2+E_3A_3B_4^2+E_4A_4B_5^2)\overline{A}_{ni}k_{ni}^{*3}e^{k_{ni}^{*}L_1}
+ \sum_{i=1}^{8} (E_1A_1B_2B_7 - E_3A_3B_4B_8
+ E_4A_4B_2B_5)h_{c1}f_{ni}\overline{A}_{ni}k_{ni}^{*2}\, e^{k_{ni}^{*}L_1}
+ \sum_{i=1}^{8} (E_4A_4B_5B_6 - E_1A_1B_3B_7
- E_3A_3B_3B_4)h_{c2}g_{ni}\overline{A}_{ni}k_{ni}^{*2}\, e^{k_{ni}^{*}L_1} - S_2 = 0 \tag{22}
$$

The shear force balance at the interface, as viewed from Section 2, leads to:

$$
D_{t1}\left[\hat{W}_{2n}(L_1)\left(\frac{\omega_n}{L_2}\right)^3 V\left(\frac{\omega_n L_1}{L_2}\right) + \hat{W}_{2n}'(L_1)\left(\frac{\omega_n}{L_2}\right)^2 S\left(\frac{\omega_n L_1}{L_2}\right) + \hat{W}_{2n}''(L_1)\left(\frac{\omega_n}{L_2}\right) T\left(\frac{\omega_n L_1}{L_2}\right) + \hat{W}_{2n}'''(L_1)\, U\left(\frac{\omega_n L_1}{L_2}\right)\right] + S_2 = 0 \tag{23}
$$

The equilibrium of the moment generated by the shear of each of the two damping layers is given by the following two equations:

$$
\sum_{i=1}^{8} (E_3A_3B_4B_8-E_1A_1B_2B_7
-E_4A_4B_2B_5)h_{c1}\overline{A}_{ni}k_{ni}^{*2}e^{k_{ni}^{*}L_1}
- \sum_{i=1}^{8} (B_2^2E_1A_1+E_3A_3B_8^2
+E_4A_4B_2^2)h_{c1}^2f_{ni}\overline{A}_{ni}k_{ni}^{*}\, e^{k_{ni}^{*}L_1}
+ \sum_{i=1}^{8} (E_1A_1B_2B_3-E_3A_3B_3B_8
- E_4A_4B_2B_6)h_{c1}h_{c2}g_{ni}\overline{A}_{ni}k_{ni}^{*}\, e^{k_{ni}^{*}L_1} = 0 \tag{24}
$$

$$
\sum_{i=1}^{8} (E_1A_1B_3B_7+E_3A_3B_3B_4-E_4A_4B_5B_6)h_{c2}\overline{A}_{ni}k_{ni}^{*2}\, e^{k_{ni}^{*}L_1}
- \sum_{i=1}^{8} (E_1A_1B_2B_3-E_3A_3B_3B_8
+ E_4A_4B_2B_6)h_{c1}h_{c2}f_{ni}\overline{A}_{ni}k_{ni}^{*}\, e^{k_{ni}^{*}L_1}
- \sum_{i=1}^{8} (E_1A_1B_3^2+E_3A_3B_3^2
+E_4A_4B_6^2)h_{c2}^2\, g_{ni}\overline{A}_{ni}k_{ni}^{*}\, e^{k_{ni}^{*}L_1} = 0 \tag{25}
$$

at x = L:

$$
\hat{W}_{2n}(L_1)\left(\frac{\omega_n}{L_2}\right)^2 S\left(\frac{\omega_n L}{L_2}\right) + \hat{W}_{2n}'(L_1)\left(\frac{\omega_n}{L_2}\right) T\left(\frac{\omega_n L}{L_2}\right) + \hat{W}_{2n}''(L_1)\, U\left(\frac{\omega_n L}{L_2}\right) + \hat{W}_{2n}'''(L_1)\left(\frac{L_2}{\omega_n L}\right) V\left(\frac{\omega_n L}{L_2}\right) = 0 \tag{26}
$$

$$\hat{W}_{2n}(L_1)\left[\frac{\omega_n}{L_2}\right]^3 V\left[\frac{\omega_n L}{L_2}\right]$$
$$+ \hat{W}_{2n}'(L_1)\left[\frac{\omega_n}{L_2}\right]^2 S\left[\frac{\omega_n L}{L_2}\right]$$
$$+ \hat{W}_{2n}''(L_1)\left[\frac{\omega_n}{L_2}\right] T\left[\frac{\omega_n L}{L_2}\right] \qquad (27)$$
$$+ \hat{W}_{2n}'''(L_1)\, U\left[\frac{\omega_n L}{L_2}\right] + \frac{M_0}{m_0}\,\frac{\omega_n^4}{L_2^2} = 0$$

The other two supplementary conditions representing the equilibrium of the axial forces in the constraining layers are:

$$E_3 A_3 \frac{\partial u_3}{\partial x} = 0 \qquad (28a)$$

$$E_4 A_4 \frac{\partial u_4}{\partial x} = 0 \qquad (28b)$$

They may be written as:

$$\sum_{i=1}^{8}\left(B_4 k_{ni}^{*2} e^{k_{ni}^* L_1} - B_8 h_{c1} f_{ni} k_{ni}^* e^{k_{ni}^* L_1} - B_3 h_{c2} g_{ni} k_{ni}^* e^{k_{ni}^* L_1}\right) = 0 \qquad (29a)$$

$$\sum_{i=1}^{8}\left(B_8 k_{ni}^{*2} e^{k_{ni}^* L_1} + B_2 h_{c1} f_{ni} k_{ni}^* e^{k_{ni}^* L_1} + B_6 h_{c2} g_{ni} k_{ni}^* e^{k_{ni}^* L_1}\right) = 0 \qquad (29b)$$

We can eliminate M_2 and S_2 from the above equations by combining Eqs. (20) and (21), and Eqs. (22) and (23). Hence, we have a total of 14 equations to determine the unknowns. In the above equations, if $M_0 = 0$, then the equations reduce to the case of the double sandwich-type cantilever with no mass at the free end. Equations (13) and (15-29) are nonlinear, complex valued equations for unknowns, Ω_n^*. A numerical scheme was used to obtain the results shown below. The uncovered cantilever beam with tip mass provided starting values for the scheme. These values and the secant method were employed to solve for Ω_n^*. By assuming a complex frequency factor, p^*, the real frequency factor (or the resonant frequency), p, and the loss factor, η, of the beam are related to Ω_n^* by:

$$\left.\begin{aligned} p^* &= \Omega^* t_0 = p(1+j\eta)^{1/2} \\ p &= \Omega\, t_0 = \sqrt{Re(p^{*2})} \\ \eta &= Im(p^{*2})/Re(p^{*2}) \end{aligned}\right\} \qquad (30)$$

where $t_0 = \sqrt{mL^4/D}$; m is the mass per unit length of Section 1 and D is the effective bending stiffness of Section 1, i.e., $\sum_{i=1,3,4} (E_i h_i^3) b/12$.

NUMERICAL RESULTS

Numerical results for the various mass ratios were obtained and are displayed as graphs. The input parameters employed in the previously described numerical scheme were the material and geometrical properties, mass density, and core loss factor. The values chosen for the computation, unless stated otherwise, were $E_1=E_3=E_4=20.6\times10^{10}$ N/m^2; $G_{c1}=G_{c2}=0.42\times10^{10}$ N/m^2; L=0.5 m, h_1=0.02 m; $\rho_1=\rho_3=\rho_4=7850$ Kg/m^3; $\rho_{c1}=\rho_{c2}=3140$ Kg/m^3; and $h_3=h_4=h_{c1}=h_{c2}=1/2h_1$. It should be mentioned that the core shear modulus is assumed to be constant in the calculation. But in an actual case, the core shear modulus varies with frequency and temperature.

Figure 2 shows the lowest resonant frequency versus the mass ratio (the attached mass to mass of the beam). It can be observed that the increase in mass ratio will decrease the value of the lowest resonant frequency of the beam. Also, the increase in the damping and constraining layer length will increase the value of the resonant frequency of the beam. This latter effect is the same as that obtained in Levy and Chen. This occurs, in this case, because the constraining layer values of E_3 and E_4 are the same as E_1, and the shear moduli are much smaller in magnitude. This has the effect of moving the cantilever end of the beam to L_1, thereby shortening the effective length of the beam. This, of course, increases the resonant frequency. It must be noted that the results are highly dependent on the input data and that the figures may have different trends for other input parameters.

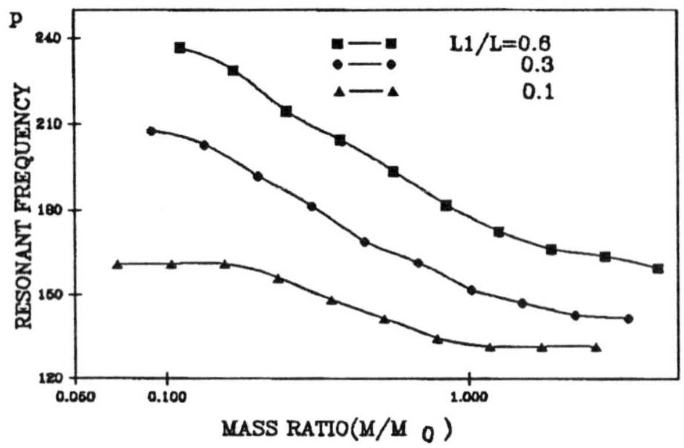

FIG. 2

VARIATIONS OF RESONANT FREQUENCY VERSUS MASS RATIO

Figure 3 represents the system loss factor versus the damping coverage. Here the core loss factor, η_c, is taken as 1 to simplify the calculations. It may be seen from the figure that an increase in damping coverage will increase the system loss factor. The system loss factor is related to energy dissipation. As more damping material covers the beam, more energy is dissipated, thus increasing the system loss factor.

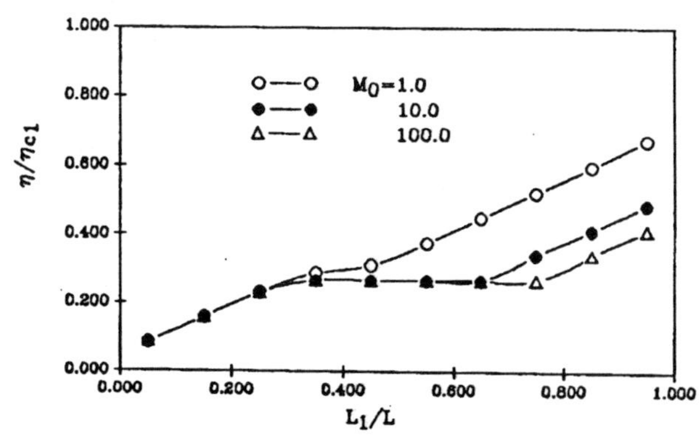

FIG. 3

VARIATION OF SYSTEM LOSS FACTOR WITH DAMPING COVERAGE AND MASS

An increase of the concentrated mass will also decrease the value of the system loss factor at constant coverage ratios. When the concentrated mass is very small, compared to the mass of the beam, the increase of L_1/L will increase the value of the system loss factor; but if the mass ratio is large, the increase of L_1/L will decrease the value of the system loss factor. This may be explained as follows: The system loss factor is related to energy loss and the energy of such a viscoelastic system is dependent on the resonant frequency and transverse displacement of the system. For small mass, the vibrations of the beam are damped as the coverage, L_1/L, is increased. For constant coverage, as mass is added to the free end of the system, the beam must execute larger displacements. Hence, adding mass represents a reduction of the damping effect of the coverage and decreasing energy loss — thus a lower system loss factor. After some point, additional damping coverage counters the effect of the increase mass. These explanations are consistent with Rao's formula for a single degree of freedom viscoelastic system (Rao, 1990).

CONCLUSIONS

In this paper, the differential equations of motion governing the partially covered, double sandwich-type cantilever with concentrated mass at the free end were obtained and employed to find the resonant frequency and loss factor of the system. Data obtained for the resonant frequency and system loss factor are found to be highly dependent on the input parameters. The following conclusions were obtained for our parameters: (1) an increase in mass ratio will decrease the resonant frequency; (2) an increase in damping coverage will increase the value of the system loss factor, but an increase of the mass ratio will cause a decrease or flatten the value of the system loss factor; (3) after some large value of the coverage, the system loss factor will increase again.

REFERENCES

Amba-Rao, C. L. and Hussaini, M. Y., 1975, "On Free Vibration of Finite Beams with Partially Fixed End Conditions Carrying Masses," *Journal of Applied Mechanics*, **42** (3), pp. 735-736.

Ditaranto, R. A., 1965, "Theory of Vibratory Bending for Elastic and Visco-Elastic Layered Finite Length Beams," Transactions ASME, *J. Applied Mechanics*, **87**, pp. 881-886.

Gurgoze, M., 1986, "On the Approximate Determination of the Fundamental Frequency of a Restrained Cantilever Beam Carrying a Tip Heavy Body," *J. Sound & Vib.*, **105**, pp. 443-449.

Laura, P. A. A., Susemihl, E. A., Pombo, J. L., Liusoni, L. E., and Gelos, R., 1977, "On the Dynamic Behavior of Structural Elements Carrying Elastically Mounted, Concentrated Masses," *Applied Acoustics*, **10**, pp. 121-145.

Laura, P. A. A., Filipich, C. P., and Cortinez, V. H., 1987, "Vibrations of Beams and Plates Carrying Concentrated Masses," *J. Sound & Vib.*, **117**, pp. 459-465.

Levy, C. and Chen, Q., "Vibration Analysis of the Partially Covered Double Sandwich-Type Cantilever Beam, Pt. I and Pt. II," submitted to the *Journal of Sound and Vibration.*

Pan, H. H., 1965, "Transverse Vibration of an Euler Beam Carrying a System of Heavy Bodies," *Journal of Applied Mechanics*, **32** (2), pp. 434-437.

Rami Reddy, K. and Amba-Rao, C. L., 1973, "Vibrations of Beams Carrying Masses-Transfer Matrices Approach," *Journal of the Aeronautical Society of India*, **25** (3) pp. 107-110.

Rao, D. K., 1978, "Frequency and Loss Factors of Sandwich Beams Under Various Boundary Conditions," *Journal of Mechanical Engineering Science*, **20**(5), pp. 271-282.

Rao, D. K., 1977, "Vibrations of Short Sandwich Beams," *J. Sound & Vib.*, **52**(2), pp. 253-263.

Rao, S. S., 1990, *Mechanical Vibrations*, Second Edition, Addison-Wesley Publishers, New York, Chapter 8.

Stokey, W. F. and Zorowski, C. F., 1959, "Normal Vibrations of a Uniform Plate Carrying Any Number of Finite Masses," *J. Applied Mechanics*, **81**, pp. 210-216.

APPENDIX I

The displacement and shear strain functions in the first section of the beam are assumed in the following form:

$$\overline{w}_{1n}(x) = \sum_{i=1}^{8} \overline{A}_{ni}\, e^{k_{ni}^{*}x}$$

$$\gamma_{c1n}(x) = \sum_{i=1}^{8} f_{ni}\, \overline{A}_{ni}\, e^{k_{ni}^{*}x}$$

$$\gamma_{c2n}(x) = \sum_{i=1}^{8} g_{ni}\, \overline{A}_{ni}\, e^{k_{ni}^{*}x},$$

where

$$f_{ni} = \frac{d_{23}d_{31} - d_{21}d_{33}}{d_{22}\, d_{33} - d_{32}d_{23}}$$

$$g_{ni} = \frac{d_{21}d_{32} - d_{22}d_{31}}{d_{22}\, d_{33} - d_{32}d_{23}}$$

$$d_{11} = -\, m\Omega_n^{*2} + (D + E_1A_1B_7^2 + E_3A_3B_4^2 + E_4A_4B_5^2)k_{ni}^{*4}$$

$$d_{12} = +\, (E_1A_1B_2B_7 - E_3A_3B_4B_8 + E_4A_4B_2B_5)h_{c1}k_{ni}^{*3}$$

$$d_{13} = -\, (E_1A_1B_3B_7 - E_3A_3B_3B_4 + E_4A_4B_5B_6)h_{c2}k_{ni}^{*3}$$

$$d_{21} = (E_1A_1B_2B_7 - E_3A_3B_4B_8 + E_4A_4B_2B_5)h_{c1}k_{ni}^{*3}$$

$$d_{22} = (B_2^2E_1A_1 + B_8^2E_3A_3 + B_2^2E_4A_4)h_{c1}^2k_{ni}^{*2} - G_{c1}A_{c1}$$

$$d_{23} = (-E_1A_1B_2B_3 + E_3A_3B_3B_8 + E_4A_4B_2B_6)h_{c1}h_{c2}k_{ni}^{*2}$$

$$d_{31} = (E_4A_4B_5B_6 - E_1A_1B_3B_7 - E_3A_3B_3B_4)h_{c2}k_{ni}^{*3}$$

$$d_{32} = (E_3A_3B_3B_8 - E_1A_1B_2B_3 + E_4A_4B_2B_6)h_{c1}h_{c2}k_{ni}^{*2}$$

$$d_{33} = (E_1A_1B_3^2 + E_3A_3B_3^2 + E_4A_4B_6^2)h_{c2}^2k_{ni}^{*2} - G_{c2}A_{c2}\ ,$$

where the E's, B's, and A's are as defined in the Nomenclature.

DE-Vol. 63, Vibrations of Mechanical Systems
and the History of Mechanical Design
ASME 1993

AN EXPERIMENTAL AND THEORETICAL STUDY OF DYNAMIC STRUCTURAL STIFFNESS IN COMPLIANT FOIL BEARINGS

C.-P. Roger Ku
Mechanical Technology Incorporated
Latham, New York

ABSTRACT

This paper describes an experimental investigation into the dynamic characteristics of corrugated foil (bump foil) strips used in compliant surface foil bearings. This study provided the first opportunity to quantify the dynamic structural stiffness of bump foil strips for a wide range of operating conditions. The experimental data were compared to results obtained by a theoretical model developed earlier, and the comparisons show very good agreement. The effects of bearing design parameters, such as static loads, dynamic displacement amplitudes, bump configurations, pivot locations, surface coatings, and lubricant were also evaluated. An understanding of the dynamic characteristics of bump foil strips resulting from this work offers designers a means for enhancing the design of high-performance compliant foil bearings.

NOMENCLATURE

E_B Bump elastic modulus
K Bump foil strip dynamic structural stiffness
K^* Dimensionless bump foil strip dynamic structural stiffness
W Bearing static load
ΔW Bearing dynamic load
W_T Unit load along transverse direction = 175 N/m (1 lb/in.)
ℓ_r Reference bump half length = 1.27 mm (0.050 in.)
m Number of bumps in a bump foil strip
s Bump pitch
t_B Bump thickness
t_{Br} Reference bump thickness = 76.2 μm (0.003 in.)
u_B Bump transverse length
Ω Dynamic displacement amplitude vibration frequency
δ Dynamic displacement amplitude of bearing pivot location
δ^* Dimensionless dynamic displacement amplitude of bearing pivot location
ν_B Poisson's ratio of bump

INTRODUCTION

Over the past 10 years, the ability of a conventional bearing to survive the rigors of today's advanced turbine engines and rocket propulsion systems has gradually declined. Extreme environments are pushing conventional liquid fluid-film bearings to, or perhaps beyond, their operating limits within high-temperature and cryogenic settings. In some turbine engines, bearing temperatures are expected to exceed the capabilities of conventional liquid lubricants completely; in rocket propulsion, bearing deficiencies persist in cryogenic turbopumps. On the other hand, compliant foil bearings, which can operate on either gas or liquid, have demonstrated good performance at elevated temperatures and high speeds, and are very attractive for use with cryogenic fluids.

The resilience offered by a compliant foil bearing stems from its construction of a smooth top foil, which provides the bearing surface, and a flexible, corrugated bump foil strip, which provides a resilient support to the top foil (see Figure 1). The bump foil strip is welded at one end to the bearing housing and is free at the other end. The advantages offered by the compliant foil bearing over conventional bearings include its adaptation to shaft misalignment, variations due to tolerance build-ups, centrifugal shaft growth, and differential thermal expansion. It has long life and reliability, higher load capacity, a lower power loss, and superior rotordynamic characteristics (Gary et al., 1981).

Existing analytical models of foil bearings are not as developed, however, as those of conventional-type fluid-film bearings. Walowit et al. (1973) first introduced a theoretical model to determine the static structural stiffness of the resilient support. This model assumed that bumps do not interact with each other, thus neglecting local interactive forces between bumps as well as friction forces between the top foil and the bumps. According to this model, when the top foil is loaded,

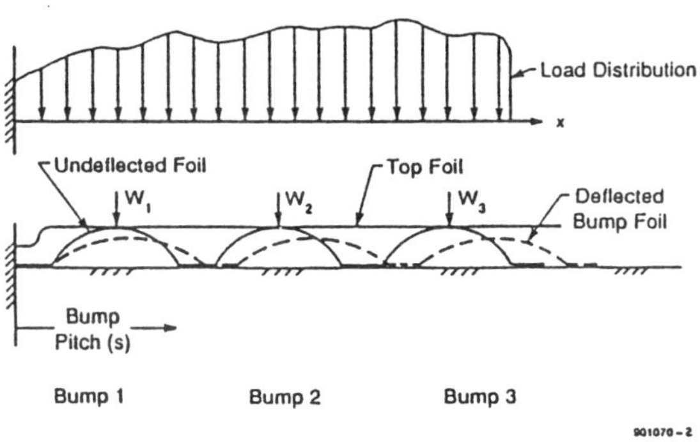

FIGURE 1 BUMP FOIL STRIP WITH AND WITHOUT APPLIED LOAD.

each bump has identical deflection and static stiffness.

Ku and Heshmat (1992) recently developed a theoretical model to investigate the mechanism of deformation of a bump foil strip used in thrust foil bearings. This work was a first step toward understanding the relationship between frictional and local interacting forces in bump foil strips. In this model, the friction forces between top foil and bumps, the friction forces between housing and bumps, and the local interactive forces between bumps are taken into consideration. These researchers predicted that the bumps near the fixed end would have higher static stiffness than the bumps near the free end, and that the load distribution profile would have a great effect on bump local static stiffness.

More recently, Ku and Heshmat (1993b,c) extended their model to predict the dynamic structural stiffness and equivalent viscous damping coefficients of the resilient support, the bump foil strip, in a journal bearing or damper. Stiffness is calculated based on the perturbation of the journal center with respect to its static equilibrium position. The equivalent viscous damping coefficient is determined based on the area of a closed hysteresis loop of the journal center motion. They found, theoretically, that the energy dissipated from this loop is mostly contributed by the frictional motion between contact surfaces. With the introduction of this enhanced model, the analytical tools are now available for the design of compliant foil bearings.

In a follow-up experimental investigation, Ku and Heshmat (1993a) studied the two-dimensional deflections of bump foil strips via an optical tracking system to verify the feasibility of the theoretical model. They measured the static stiffness of bump foil strips and reported that the stiffness is load and/or amplitude dependent.

In this paper, the experimental study has been extended to quantify the dynamic structural stiffness of bump foil strips. The results are compared to analytical predictions obtained from the previously developed theoretical model. The effects of static load, dynamic displacement amplitude, bump configurations, pivot locations, surface coatings and lubricant are also investigated.

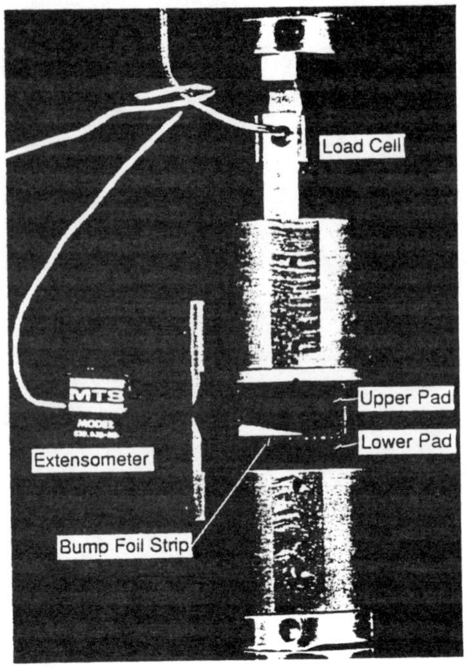

FIGURE 2 BUMP FOIL ASSEMBLY ON TEST APPARATUS.

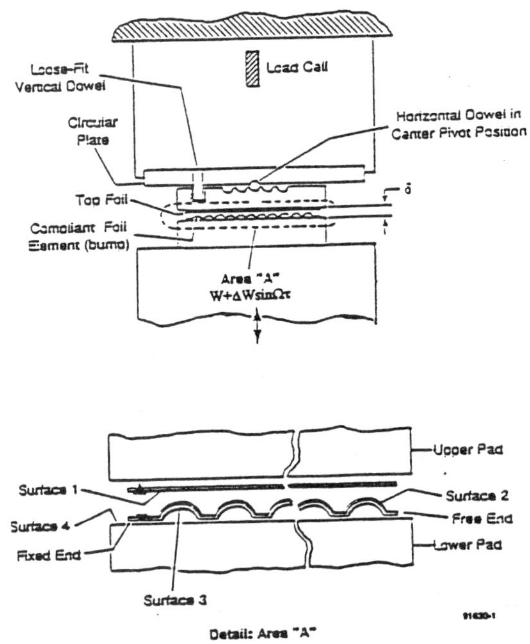

FIGURE 3 BUMP FOIL PIVOT LOCATIONS AND IDENTIFICATION OF SURFACE CONTACTS.

TABLE 1 CONFIGURATIONS AND SURFACE COATINGS OF TESTED BUMPS

#	Pitch	Thickness	Height	Material Combination		Lubricant
	mm	μm	mm	Surfaces 1 & 2	Surfaces 3 & 4	
1	4.572	76.2	0.4953	Ni-Ni	Ni-Steel	Air
2	4.572	76.2	0.4953	Ni-Ni	Ni-Steel	Oil
3	4.572	76.2	0.4953	Ni-Ni	Cu-Cu*	Air
4	4.572	76.2	0.4953	Ni-Ni	Cu-Cu*	Oil
5	4.572	76.2	0.4953	Cu-Cu*	Ni-Steel	Air
6	4.572	76.2	0.4953	Tef-Tef#	Ni-Steel	Air
7	4.572	76.2	0.4953	Ni-Ni	Tef-Tef#	Air
8	4.572	63.5	0.4890	Ni-Ni	Ni-Steel	Air
9	4.572	63.5	0.4890	Ni-Ni	Ni-Steel	Oil
10	4.191	76.2	0.4699	Ni-Ni	Ni-Steel	Air

* INC-X750 Coated with Cu
\# INC-X750 Coated with PTFE

TABLE 2 TEST PARAMETERS

Parameter	Symbol	English Unit	SI Unit
Static Load	W	20, 30 lb	90, 145 N
Perturbation Amplitude (Half)	δ	~0.0001-0.0002 in.	~2.5-5.0 μm
Perturbation Frequency	Ω	1 Hz	
Number of Bumps	m	6	---
Bump Elastic Modulus	E_B	3.0×10^7 psi	2.07×10^5 MPa
Bump Poisson's Ratio	ν_B	0.25	---
Bump Transverse Length	u_B	0.95 in.	24.1 mm

EXPERIMENTAL SETUP

The test apparatus and bump foil assembly are shown in Figures 2 and 3. Each pad is 41.1 mm by 24.6 mm and made of steel. The lower pad acts as a housing and the upper pad is supported by a bump foil strip. The bump foil strip is welded at one end of the lower pad; a smooth top foil is welded on the lower surface of the upper pad. As the lower pad moves or vibrates in the vertical direction, the bumps are deformed in both vertical and horizontal directions. The vertical deflections of the bumps determine load capacity, or stiffness, and the horizontal motions of the bumps yield damping.

A horizontal hard dowel, which acts as a pivot, is placed between a circular plate and the upper pad. The pivot can be located in any one of five grooves that are spaced 3.96 mm apart on the top surface of the upper pad. A vertical dowel, inserted tightly into the circular plate and loosely into the upper pad, prevents the horizontal motion of the upper pad. When the pivot location is changed, the upper and lower pads move as a unit.

The lower pad is mounted on the lower adaptor by two roll pins. The circular plate with a central groove is fixed on the upper adaptor. The upper and lower adaptors are mounted on an MTS hydraulic force control system. An extendsometer, which has a sensitivity of 25 μm/V, controls the relative distance between the adaptors. A force transducer (44.5 N/V) measures the load capacity of the bump foil strip. The optical tracking system used for the static deformation study (Ku & Heshmat, 1993a) was removed. An Ono-Sokki two-channel analyzer and a plotter record the test data.

The bump foil strips and smooth top foils were made of Inconel X750; all the other parts were made of steel. Each bump foil strips is 24.1 mm wide and has six bumps. The surfaces of some of the foils were coated with either Teflon or copper. For the bump foil strip with one of these coatings, another smooth foil with the same coating was welded between the bump foil strip and the upper surface of the lower pad (see Figure 3, Surface 4). In this way, all contact surfaces had the same coating. For each coating, both dry contact surfaces and surfaces lubricated with light turbine oil were tested. Table 1 shows the configurations and coatings of the tested bump foils.

The detailed experimental procedures used to conduct the static load test were described in Ku & Heshmat (1993a). After a static load was applied, a dynamic force or displacement was imposed. In the current experiments, a dynamic displacement of the vertical deformation of the bump foil strip at the pivot location was controlled by the extendsometer, and the corresponding dynamic force was recorded. Dynamic load-

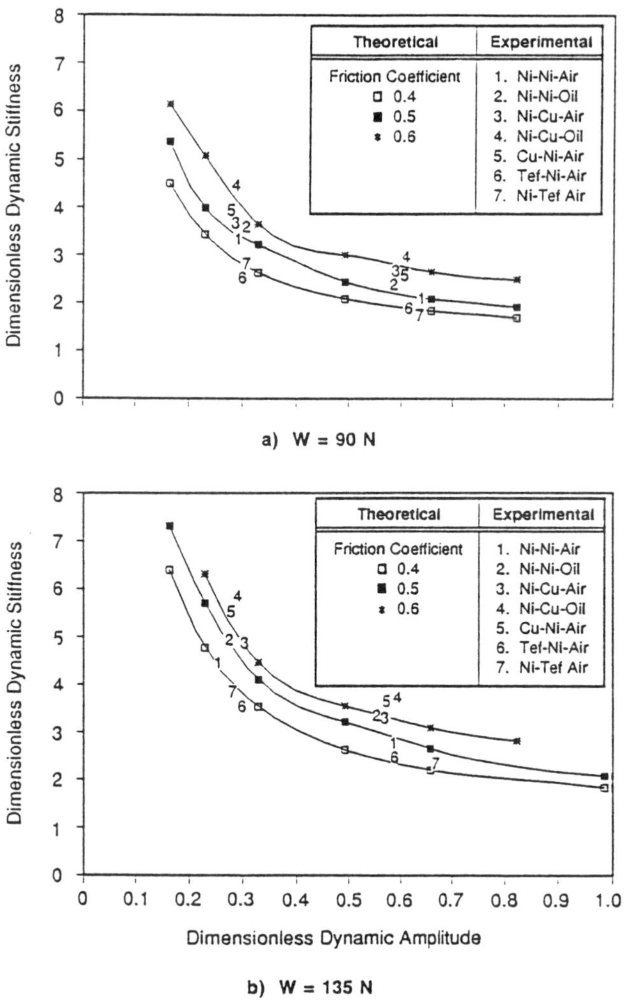

FIGURE 4 EFFECT OF SURFACE COATINGS (s=4.6 mm, t_B=76 μm, CENTER PIVOT).

deflection curves (hysteresis loops) were plotted with respect to the static equilibrium positions. For a specific static load and dynamic displacement, the dynamic structural stiffness of the bump foil strip was calculated by taking the net force divided by the net displacement amplitude of the dynamic load-deflection curve.

For each bump foil strip specimen listed in Table 1, three most centrally located pivot positions were evaluated (see Figure 3). At each pivot location, two static loads (90 and 135 N) and two dynamic displacement amplitudes (~ 2.5 and 5.0 μm) at 1 Hz frequency were tested, respectively. The test parameters are shown in Table 2.

RESULTS AND DISCUSSION

The experimental results are displayed in Figures 4 through 6. The resulting data points are represented by numbers, which are identical to the test number shown in the first column of Table 1. The dimensionless variables shown in the figures are defined as follows:

Dimensionless displacement amplitude,

$$\delta^* = \delta \frac{E_B}{2W_T(1-\nu_B^2)}\left(\frac{t_{Br}}{\ell_r}\right)^3 \tag{1}$$

Dimensionless structural stiffness,

$$K^* = K\frac{2(1-\nu_B^2)}{mu_B E_B}\left(\frac{\ell_r}{t_{Br}}\right)^3 \tag{2}$$

where E_B, ν_B, t_{Br}, ℓ_r, W_T, and m are known parameters (see Table 2 and Nomenclature).

The analytical results were calculated by using the previously developed theoretical model (Ku & Heshmat, 1992, 1993a,b), in which the dynamic structural stiffness was determined based on the perturbation of the pivot location with respect to its static equilibrium position. Although it is very difficult to measure the friction coefficients accurately for different surface coatings, these coefficients are required as input parameters in the theoretical model. Therefore, the sum of the friction coefficients between Surfaces 1 and 2 and Surfaces 3 and 4 (Figure 3) was treated as unknown parameters in the current study, and these coefficients were determined empirically. The results of dynamic stiffness for different total friction coefficients were plotted by the solid lines in Figures 4 through 6.

Figure 4 displays the effect of surface coatings and lubricant on dynamic structural stiffness at different static loads and dynamic displacement amplitudes. For each of the seven cases (1-7), the bump had a 4.6-mm pitch and a 76-μm thickness, and the pivot was in the center position. Note that the coating with a higher friction coefficient provides higher dynamic structural stiffness. Also, the bumps with either Ni-Cu or Cu-Ni coatings (the first chemical element represents the coating of Surfaces 1 and 2, and the second one represents the coating of Surfaces 3 and 4, see Figure 3) have roughly the same stiffness, implies that the dynamic structural stiffness is primarily affected by the total friction coefficients of the contact surfaces, i.e., between the bump and the top foil and between the bump and lower pad. Switching the order of the surface coatings, for example, from Ni-Tef to Tef-Ni, has less effect on the dynamic structural stiffness than does changing one of the surface coatings , such as from Ni-Tef to Ni-Cu. The earlier theoretical calculations and experimental measurements of the static structural stiffness resulted in similar conclusions (Ku & Heshmat, 1992, 1993a).

In comparing the experimental data with the theoretical results, the total friction coefficients were determined empirically by matching the values of dynamic structural stiffness. From Figure 4 we observed that the Ni-Tef (or Tef-Ni) surface coating has a total friction coefficient near 0.4; the Ni-Ni surface coating has a total friction coefficient near 0.5; and the Ni-Cu (or Cu-Ni) surface coating has a total friction coefficient near 0.6.

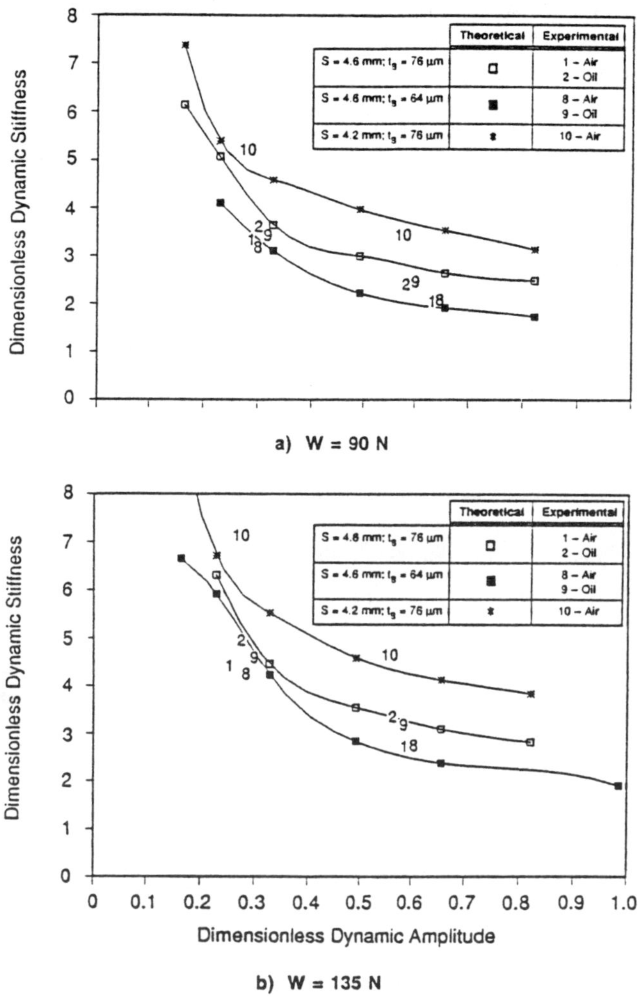

FIGURE 5 EFFECT OF BUMP CONFIGURATIONS (NI-NI SURFACE COATING, CENTER PIVOT).

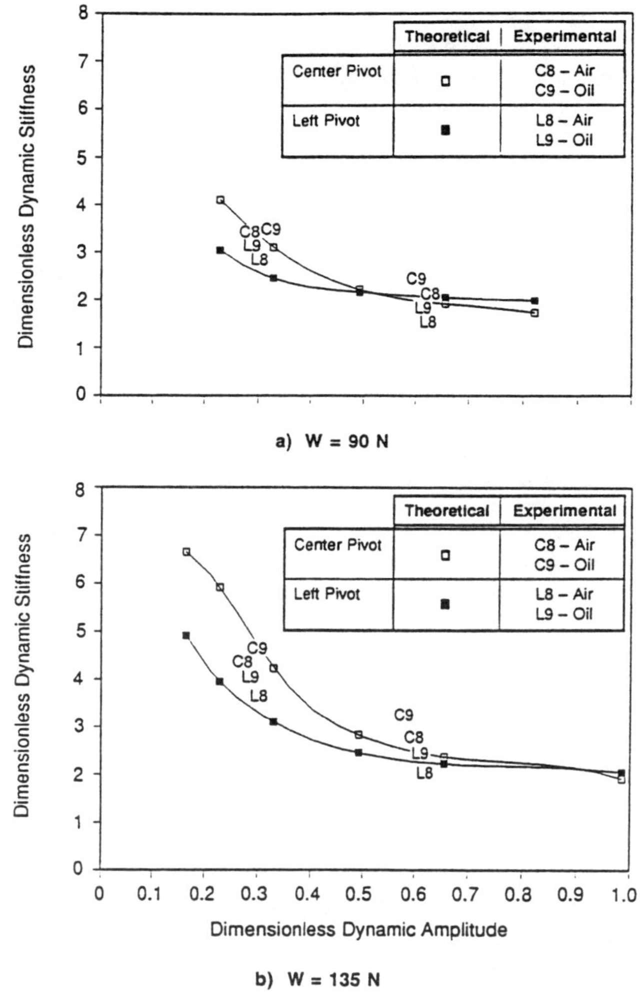

FIGURE 6 EFFECT OF PIVOT LOCATIONS (s=4.6 mm, t_B=76 μm, NI-NI SURFACE COATING).

Figure 5 displays the effect of three bump configurations on dynamic structural stiffness. For each configuration, the pivot was located in the center for a Ni-Ni surface coating with (cases 2 and 9) and without (cases 1, 8, and 10) lubricant. The theoretical results were calculated with total friction coefficients 0.6. At both static loads and dynamic displacement amplitudes, increasing the bump thickness and/or pitch would increase the values of dynamic structural stiffness. Note that changing the pitch had a greater effect than changing the bump thickness.

The effect of pivot location on dynamic structural stiffness is shown in Figure 6. For each pivot location, center or left, the bump had a 4.6-mm pitch and a 76-μm thickness, and a Ni-Ni surface coating with (cases C9 and L9) and without (cases C8 and L8) lubricant. The center pivot tends to provide a little more dynamic structural stiffness than the left pivot for both static loads.

The effect of lubricant on dynamic structural stiffness is also shown in Figures 4 through 6. The addition of oil to the surfaces coated with nickel (cases 2 and 9) and copper (case 4) increases the dynamic stiffness for all test conditions. This phenomenon may be due to the fact that the surface contact occurred in the boundary lubrication regime, therefore, the adhesive effect resulted in an increase of the friction coefficient (Israelachvili, 1992).

CONCLUSIONS

An experimental study was performed to quantify the dynamic structural stiffness of bump foil strips used in foil bearings for a wide range of operating conditions. The results were compared to the analytical predictions by using the previously developed theoretical model. The friction coefficients between the contact surfaces for different surface coatings were determined empirically by matching the values of dynamic structural stiffness. The effects of bump configurations, pivot locations, surface coatings, lubricant, static load, and dynamic displacement amplitude on the bump foil strip dynamic structural stiffness were also evaluated.

It was shown that the analytical predictions and experimental measurements agreed well. As the static load increases or dynamic displacement amplitude decreases, the dynamic structural stiffness increases for all the test conditions. The bearing designer may use the coating with a higher friction coefficient, add the lubricant to the surfaces, increase the bump thickness or decrease of the bump pitch, and choose the center pivot location to achieve higher dynamic structural stiffness.

An understanding of the dynamic characteristics of bump foil strips resulting from this work offers designers a means for enhancing the design of high-performance compliant foil bearings. Recently, a rotor with 35 mm (1.375 in.) diameter and 15.2 N (3.41 lb) weight has been operated with two air foil bearings to 132,000 rpm successfully. These bearings have demonstrated a load capacity to 0.67 MPa (97 psi) (Heshmat, 1993).

ACKNOWLEDGMENTS

The work was principally supported by Mechanical Technology Incorporated. The author express his appreciation to the organization for their support. The author would also like to thank his colleagues, Mr. M. Tomaszewski, for his help in preparing the test facilities, and Dr. H. Heshmat for his many valuable suggestions and discussions.

REFERENCES

Gray, S., Heshmat, H., and Bhushan, B., 1981, "Technology Progress on Compliant Foil Air Bearing Systems for Commercial Applications," *8th International Gas Bearing Symposium*, pp. 69-97.

Heshmat, H., 1993, "The Advancement in Performence of Aerodynamic Foil Journal Bearings: High Speed and Load Capability," submitted to *ASME/STLE Joint Tribology Conference*, Oct. 24-27, New Orleans, Louisiana.

Israelachvili, J. N., 1992, "Adhesion, Friction and Lubrication of Molecularly Smooth Surfaces," Fundamentals of Friction: Macroscopic and Microscopic Processes, Edited by I. L. Singer and H. M. Pollock, Kluwer Academic Publishers, pp. 351-385.

Ku, C.-P. R., and Heshmat, H., 1992, "Compliant Foil Bearing Structural Stiffness Analysis Part I: Theoretical Model - Including Strip and Variable Bump Foil Geometry," ASME, *Journal of Tribology* Vol. 114, No. 2, pp. 394-400.

Ku, C.-P. R., and Heshmat, H., 1993a, "Compliant Foil Bearing Structural Stiffness Analysis Part II: Experimental Investigation," accepted for publishing in *ASME, Journal of Tribology*, paper No. 92-Trib-6.

Ku, C.-P. R., and Heshmat, H., 1993b, "Structural Stiffness and Coulomb Damping in Compliant Foil Journal Bearings: Theoretical Considerations," accepted for publication in *STLE Tribology Transactions* and presentation at *STLE Annual Meeting*, May 17-20, 1993, Calgary, Canada.

Ku, C.-P. R., and Heshmat, H., 1993c, "Structural Stiffness and Coulomb Damping in Compliant Foil Journal Bearings: Parametric Studies," accepted for publication in *STLE Tribology Transactions* and presentation at *STLE Annual Meeting*, May 17-20, 1993, Calgary, Canada.

Walowit, J. A., Murray, S. F., McCabe, J., Arwas, E. B., and Moyer, T., 1973, "Gas Lubricated Foil Bearing Technology Development for Propulsion and Power Systems," WPAFB Technical Report AFAPL-TR-73-92, Air Force Aero Propulsion Laboratory, Wright-Patterson Air Force Base, Ohio.

DE-Vol. 63, Vibrations of Mechanical Systems
and the History of Mechanical Design
ASME 1993

INTERVAL ANALYSIS OF VIBRATING SYSTEMS

Andrew D. Dimarogonas
Department of Mechanical Engineering
Washington University
Saint Louis, Missouri

ABSTRACT

Interval calculus is a tool to evaluate a mathematical expression for ranges of values of its parameters. The basic mathematical operations are defined in the interval algebra.

Vibrating systems having system parameters, initial conditions or forcing functions defined as intervals rather than single-valued quantities have been modeled with interval calculus.

The eigenvalue problem for determining natural frequencies and vibrating modes of a system in the case of system matrix elements given as a range of values cannot be solved by exhaustion due to the prohibitively large number of solutions of the point-number eigenvalue problem. Eigenvalue solution with interval evaluation of the commonly used numerical techniques is not feasible because if they are applied directly, the solution intervals diverge. An optimization technique was used to obtain the minimum-radius intervals of the solution. To assure monotonicity and absolute inclusion, necessary for convergence to the exact interval, a converging interval halving sequence was developed. For numerical tests of the method a Monte Carlo solution was developed. Results showed that interval analysis can predict the range of the eigenvalues with sufficient accuracy.

The response of a linear system to general excitation for interval matrices of the system parameters cannot be found with interval evaluation of the commonly used numerical techniques because if they are applied directly, the solution intervals diverge. Interval modal analysis and the interval solution of the eigenvalue problem were developed.

An application is presented for the dynamic response of a rotor with interval bearing properties.

INTRODUCTION

Archimedes (c. 287-212 BC) defined the irrational number π by way of an interval $3\ 10/71 < \pi < 3\ 1/7$ which he found by approximating the circle with the inscribed and the circumscribed 96-side regular polygons (Archimedes 287-220 B.C.). The concept of a function having values which are bounded between limits was discussed by W.H. Young (1908). The concept of operations with a set of multi-valued numbers was introduced by R.C. Young (1931) who developed a formal algebra of multivalued numbers. The special case of closed intervals was discussed by Dwyer (1951). Interval mathematics was further developed by Sunaga (1958), who introduced the term *Interval Algebra*, and Moore (1966), who introduced *Interval Analysis*, interval vectors and matrices and the first non-trivial applications, such as the initial value problem for ordinary differential equations with interval initial conditions.

Systematic treatment of interval vectors and matrices was introduced by Apostolatos and Kulich (1968). Alefeld and Herzberger (1983) presented an extensive treatment of interval linear and nonlinear algebraic equations. Kulish and Miranker (1981) focused on interval arithmetic applications to computer arithmetic, which is by its nature an interval arithmetic. In a recent monograph, Neumaier (1990) reviewed interval methods for systems of equations.

A substantial number of engineering vibration problems are indeed interval problems. All rotor dynamics analyses are performed with single-valued bearing properties. It is well-known, however, that the bearing properties change substantially with temperature, which is controlled in machinery within relatively broad intervals. Since the viscosity depends on temperature with a double exponential law, even small temperature changes result in substantial changes in the bearing properties. Manufacturing tolerances introduce another interval parameter, the bearing clearance, which widens further the bearing parameters.

Another source of interval values for system properties are the boundary conditions. Simple support or fixed ends are idealizations of the boundary conditions which are far from being accurate. An interval evaluation of the system response would be more rational.

Statistical methods have been used to address the problem of uncertainty in the operating conditions or the system parameters. This, however, is valid under very strong restrictions on the uncertainties considered. For problems, in particular, where the distribution of a property is unpredictable or changes drastically during the process, interval analysis can provide a rational method to determine a conservative, and under certain conditions exact, interval for the result.

INTERVAL ALGEBRA CONVENTIONS

In the following, the field of real numbers is denoted by $\Re$ and its members by lower case letters. A subset of $\Re$ of the form

$$A = [a_1, a_2] = \{t | a_1 \le t \le a_2, a_1, a_2 \in \Re\} \quad (1)$$

will be called a (closed) interval and a_1 is the lower bound and a_2 the upper bound, denoted by inf(A) and sup(A), respectively. The mid-point is defined as mid(A) = $(a_1 + a_2)/2$ and the radius as rad(A) = $(a_2 - a_1)/2$. The set of closed real intervals will be denoted by $I(\Re)$ and its members by uppercase letters. For clarity, matrices and vectors will be denoted by boldface letters, i.e. **A**, **B**, **a**, **b**, and interval matrices and vectors, defined as matrices and vectors with interval number elements, by boldface and underlined letters, i.e. $\underline{\mathbf{A}}$, $\underline{\mathbf{B}}$, $\underline{\mathbf{a}}$, $\underline{\mathbf{b}}$.

Two sets $A = [a_1, a_2]$ and $B = [b_1, b_2]$ are equal iff $a_1 = b_1$, $a_2 = b_2$. An interval of zero width [a, a] will be called point interval and it is a regular real number. The following set of binary operations is defined (Moore 1966):

$$A + B = [a_1 + b_1, a_2 + b_2] \quad (2)$$

$$A - B = [a_1 - b_1, a_2 - b_2] \quad (3)$$

$$AB = [\min\{a_1b_1, a_2b_2, a_1b_2, a_2b_1\}, \max\{a_1b_1, a_2b_2, a_1b_2, a_2b_1\}] \quad (4)$$

$$A : B = [a_1, a_2]\,[1/b_2, 1/b_1] \quad (5)$$

It is apparent that division is not defined if $0 \in [b_1, b_2]$ and that $[a_1, a_2]c = [ca_1, ca_2]$.

If f(x) is a continuous unary operation on $\Re$, then the interval evaluation of the function is defined over the interval number X as

$$F(X) = [\inf_{x \notin X}\{f(x)\}, \sup_{x \notin X}\{f(x)\}] \quad (6)$$

The definition of elementary operations follows immediately:

$$F(X) = [f\{\inf(X)\}, f\{\sup(X)\}] \quad (7)$$

where $f(x) \in \{e^x, \log x, \sqrt{x}, x^{2n+1}\}$ are monotonic functions of x, meaning that if $x_1 < x_2$ then $f(x_1) < f(x_2)$ for increasing and $f(x_1) > f(x_2)$ for decreasing monotonic function, respectively. Monotonicity is very important in interval arithmetic because, if f(x) is not monotonic, the interval of the function is not the function of the interval because a value within the interval yields the highest or lowest value of the function in the interval, depending on the existence of a maximum or a minimum value of the function within the interval. For example, $SIN[0,2\pi]$ = [1, 1], while $\sin(0) = 0$ and $\sin(2\pi) = 0$. Therefore, if n is an integer,

$$SIN(X) = [\sin\{\inf(X)\}, \sin\{\sup(X)\}] \quad \text{if } (2n+1)\pi/2 \notin X,\ (2n+3)\pi/2 \notin X,$$

$$SIN(X) = [1, \sin\{\sup(X)\}] \quad \text{if } (2n+1)\pi/2 \notin X,\ (2n+3)\pi/2 \in X \quad (7a)$$

$$SIN(X) = [\sin\{\inf(X)\}, 1] \quad \text{if } (2n+1)\pi/2 \in X,\ (2n+3)\pi/2 \notin X,$$

$$SINX = [1, 1] \quad \text{if } (2n+1)\pi/2 \in X,\ (2n+3)\pi/2 \in X.$$

Similar expressions can be developed for the other harmonic functions. Moreover,

$$X^{2n} = [0, |X|^{2n}] \text{ if } 0 \in X, \quad (7b)$$

$$X^{2n} = [\{\inf(|X|)\}^{2n}, \{\sup(|X|)\}^{2n}] \text{ if } 0 \notin X.$$

While commutativity and associativity are preserved in the interval algebra, distributivity is not

$$A(B + C) \neq AB + AC. \quad (8)$$

However, the following subdistributivity, or inclusion, rule is true (Sunaga 1958):

$$A(B + C) \subseteq AB + AC \quad (9)$$

The subdistributivity law, or inclusion theorem, is of fundamental importance to the interval calculus because it proves that interval arithmetic operations always yield an upper bound for the interval of the function, thus they provide a conservative estimate for it. According with the inclusion theorem, equation 9, the interval of a function is included in its interval evaluation:

$$w[f(x), x \in X] = [\inf_{x \in X}\{f(x)\}, \sup_{x \in X}\{f(x)] \subseteq F(X) \quad (10)$$

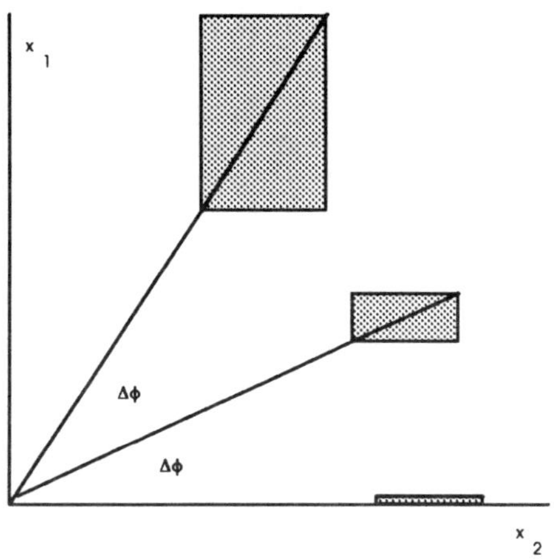

Figure 1: Solution of equations 13

An interval numerical procedure is convergent if the lower and upper limits of the solution converge to the lower and upper limits of the true solution. In interval analysis, one is satisfied many times with a non-convergent algorithm, since the subdistributivity secures inclusion, provided that the interval evaluated solution is not very far from the true solution.

INTERVAL ANALYSIS OF NATURAL VIBRATION

Simple vibration problems (1 or 2 degrees-of-freedom systems) can be solved with direct interval evaluation of the response function. For example, the general solution for a single-degree-of-freedom linear system with interval harmonic excitation $FCOS\Omega t$ and interval initial conditions X_0, V_0, is

$$X(t) = E^{-(C/2M)t}\,(X_oCOS\Omega_dt + \{X_oC/M + V_o/2\Omega_d\}\,SIN\Omega_dt)]$$

$$+ F/[(-M\Omega^2 + K)^2+(\Omega C)^2]^{1/2}\,SIN(\Omega t\text{-}\Phi) \qquad (11)$$

where $TAN\Phi = \Omega C/(-M\Omega^2+K)$,

$$\Omega_d = [K/M - (C/2M)^2]^{1/2}$$

where M, C and K are the interval system properties, F the interval excitation, Ω the interval excitation frequency and all operations are interval operations. One has to be careful in evaluating the function $(-M\Omega^2 + K)^2+(\Omega C)^2$ because its interval evaluation is correct if the function is monotonic in respect to the interval parameters. This is true for Ω, for example, iff $\Omega \notin [(K/M)(1 - 2Z^2)]^{1/2}$, where $Z^2 = C/4KM$.

After one period of the natural vibration and of the forced vibration has elapsed, SIN and COS functions reach the limit interval U = [-1, 1] and the interval response is

$$X(t) = UE^{-(C/2M)t}\,[X_o + (X_oC/M + 2V_o/2\Omega_d)] +$$

$$FU/[(-M\Omega^2 + K)^2+(\Omega C)^2]^{1/2} \qquad (12)$$

The solution of this problem is simple enough to be obtained with direct methods, though it is much more convenient to obtain it through interval arithmetic. More complex systems make direct evaluation of the interval of the response cumbersome. For reasons which became apparent already with the notice of the effect of the lack of monotonicity of the response function on the interval evaluation of the response function, direct utilization of numerical methods becomes, in general, futile. The reason is the inclusion theorem: every numerical step might lead to a grater interval, thus to a diverging process. This will become apparent with the following example:

Assume a single degree of freedom linear system expressed in normal form as

$$x_1{}^\circ = x_2$$
$$x_2{}^\circ = -x_1 \qquad (13)$$

The solution of these equations for initial conditions X_{20}= [1, 1+δ] and x_{10}= 0 is obviously X_2 = [1, 1+δ]cost, X_1 = -[1, 1+δ]sint. The development of the interval of the solution, if computed in steps of the angle of rotation $\delta\phi$, is shown in figure 1.

As the vector of the response rotates, the interval, which is evaluated always along the axes, increases at every step. The procedure clearly diverges.

Assume now another linear system

$$x_1{}^\circ = x_2$$
$$x_2{}^\circ = x_1 \qquad (14)$$

The solution of these equations for initial conditions X_{20}= [1, 1+δ] and x_{10} = 0 is obviously X_2 = [1, 1+δ]e^t, X_1 = 0. The development of the interval of the solution, if computed in steps of time δt, remains finite. The interval increases in proportion to the response and the procedure converges.

The difference in the behavior of the two systems is in the direction of monotonicity of the r.h.s. functions in the system equations (Markov and Angelov 1985). Such integration is convergent iff the r.h.s. functions are quasi-isotonous. This suggests caution in the utilization of usual numerical methods and suggests that special methods need to be developed for interval systems.

The solution of the interval eigenvalue problem is central to the interval natural and forced vibration problems and it be will considered first.

THE INTERVAL EIGENVALUE PROBLEM

Absolute bounds for the eigenvalues of an interval system were investigated by Nickel (1980). Dimarogonas (1983) and Chondros and Dimarogonas (1989) used the Rayleigh's quotient to compute the interval eigenvalues and eigenvectors for cracked structures where the cracks introduce a small range in the values of the system stiffness matrix. This method worked well, owing to the small width of the intervals involved.

Herz (1992) developed a Kharitonov-like algorithm to compute the minimal and maximal eigenvalues of an nxn symmetric interval matrix. The Herz alogorithm requires the determination of the eigenvalues of 2^n real vertex matrices. Chen (1992) used the Gershgorin Theorem to develop sufficient conditions for the stability of interval matrices.

Rohn (1990) developed a method for the computation of the interval eigenvalues of the interval real symmetric matrix $\underline{\mathbf{A}} = \{A;\ A_c - \Delta \le A \le A_c + \Delta\}$, where the radius matrix Δ is of the form $\Delta = \mathbf{q}\mathbf{p}^T$, where $\mathbf{p}$ and $\mathbf{q}$ positive vectors. Good results were obtained for small radius interval matrices.

In the general case of a system with system interval matrices $\underline{\mathbf{M}}$, $\underline{\mathbf{K}}$, which is of engineering significance, the interval eigenvalue problem is equivalent with the two optimization problems,

Find the minimum (maximum) λ
subject to $\{\lambda:\ \mathbf{Kx} = \lambda\mathbf{x},\ K \in \underline{\mathbf{K}},\ M \in \underline{\mathbf{M}}\}$
$\{\mathbf{x}:\ \mathbf{x}^T\mathbf{x} = 1\}$

This problem can be solved with a variety of methods for the general eigenvalue problem with $\underline{\mathbf{M}}$, $\underline{\mathbf{K}}$ symmetric interval matrices with some additional restrictions encountered in engineering systems. A Monte Carlo solution will be presented later.

For symmetric matrices, assuming that $\mathbf{x}_i = |\mathbf{x}_i|$ or that the eigenvectors do not change sign within the interval of the solution (Chondros and Dimarogonas 1989, Deif 1991), orthogonality demands that

$$F_i = \mathbf{x}_i^T A_c \mathbf{x}_i \pm |\mathbf{x}_i^T| \Delta A |\mathbf{x}_i| = \lambda_i. \tag{15}$$

subject to $\mathbf{x}_i^T\mathbf{x}_i = 1$ or $h_i(\mathbf{x}_i) = \mathbf{x}_i^T\mathbf{x}_i - 1 = 0$, where $\underline{A}_c$ is the mid-point matrix and ΔA the radius matrix.

Since the minimum (maximum) eigenvalues are sought, a necessary condition for minimum is given by the Karush-Kuhn-Tucker Theorem, $\Delta F_i + \lambda_i \Delta h_i = 0$, or

$$A_c\mathbf{x}_i \pm S_i \Delta A S_i \mathbf{x}_i - \lambda_i \mathbf{x}_i = 0 \tag{16}$$

where λ_i is a Langrange mutilplier. Therefore, the interval eigenvalues are:

$$\Lambda_i = [\lambda_i(A_c - S_i\Delta A S_i),\ \lambda_i(A_c + S_i \Delta A S_i)],$$
$$i = 1, 2, 3, \ldots, n \tag{17}$$

where $\mathbf{S}_i = \mathrm{diag}[\mathrm{sgn}(x_1^{\,i}), \ldots, \mathrm{sgn}(x_n^{\,i})]$ (this notation is due to Reif, 1991), and the eigenvectors $x_j^{\,i} \neq 0$. There is a restriction due to the absolute values in the objective function: within the interval of the eigenvalues, the sign matrix $\mathbf{S}_i$ of the eigenvectors should not change, in which case the bounds found for the eigenvalues are exact. This is a severe restriction but not a dangerous one because the respective eigenvectors can be computed and thus the validity of the hypothesis of no sign change can be verified *a posteriori*. To this end, the eigenvalues computed with equations 17 are used to determine the eigenvectors

$$(\underline{A} - I\lambda_i)\underline{X}_i = \underline{0} \tag{18}$$

with a unit interval value [1, 1] assigned to the last element of the eigenvector $\underline{\mathbf{X}}_i$ and solving a system of linear interval equations with the Gauss elimination method which is known (5-10) to be stable (convergent) as it is applied to interval matrices and vectors. In fact, a simple conditioning method improved convergence and reduced the width of the resulting interval solution vectors:

If the n-1 interval equations for the eigenvectors are of the form

$$\underline{\mathbf{B}}\underline{\mathbf{Y}}_i = \underline{\mathbf{C}} \tag{19}$$

after multiplication with the inverse of the mid-point matrix

$$A_c^{-1}\underline{\mathbf{B}}\underline{\mathbf{Y}}_i = A_c^{-1}\underline{\mathbf{C}} \tag{20}$$

yields a coefficient matrix with dominate diagonal elements and facilitates the solution of the system of interval linear equations with a minimum interval width.

For the general eigenvalue problem encountered in vibration analysis

$$(\underline{\mathbf{K}} - \Lambda\underline{\mathbf{M}})\underline{\mathbf{X}} = 0 \tag{21}$$

the usual form of the stiffness and mass matrices $\underline{\mathbf{K}}$ and $\underline{\mathbf{M}}$ (the first positive definite or semidefinite, the second positive definite) yields real eigenvalues and the problem can be transformed into the problem of finding the eigenvalues of a real symmetric matrix (Dimarogonas, 1992). If interval matrices are involved, equation 17 is not valid and the interval eigenvalues will be, with a similar reasoning (Deif, 1991),

$$\Lambda_i\ [\Lambda_i(A_c - \Delta A \circ S^i),\ \Lambda_i(A_c + \Delta A \circ \underline{S}^i)],$$
$$i = 1, 2, 3, \ldots, n \tag{22}$$

where the symbol $\circ$ denotes component-wise matrix multiplication, the matrix $\mathbf{S}^i$ has components $s^i_{kj} = \mathrm{sgn}(y^i_k x^i_j)$, $\mathbf{x}^i$ is the eigenvector i of the mid-point matrix A_c and $\mathbf{y}^i$ the reciprocal eigenvector to $\mathbf{x}^i$, i.e. $\mathbf{y}^i \circ \mathbf{x}^i = 1$. Again, equation 22 gives exact intervals of the eigenvalues provided that the components of the sign matrix s^i_{kj} do not change in the interval. If the eigenvectors change sign in their interval, for either one of the equations 17 and 22, one may proceed as follows:

The interval of each interval parameter is divided into two equal sub-intervals. 2^n interval eigenvalue problems are then solved with equations 22 (even for a symmetric matrix $\underline{\mathbf{A}}$) with

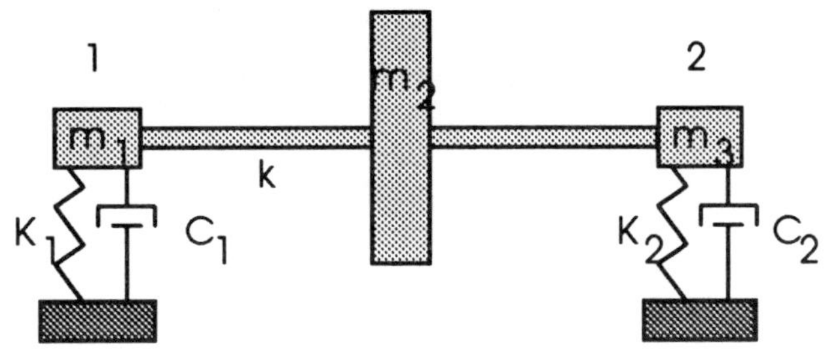

Figure 2: A 3 degree of freedom model of a turbine rotor.

all the 2^n possible combinations of the sub-intervals and the intervals of the eigenvalues will consist of the minimum and maximum values of the respective eigenvalues found in the 2^n eigenvalue solutions. The signs of the eigenvectors in each sub-interval solution are examined. If no one has changed, the procedure is terminated and the intervals found are the exact ones. If there is a change, the sub-intervals are halved and the procedure is continued. The procedure might be finite but definitely converges to the exhaustion solution, which is exact by definition.

INTERVAL SOLUTION FOR THE VIBRATION OF A 3-MASS ROTOR

A turbine rotor can be modeled by way of a three-mass lumped system m_1 ,m_2, m_3, connected with a massless shaft (Dimarogonas 1992)]. The two end masses represent the journals which are supported by linear bearings having independent interval stiffnesses K_1, K_2. The horizontal and vertical vibration are not coupled. For a numerical application: m_1=300 kg, m_2= 300 kg, m_3 = 300 kg, $K_1 = [4\text{x}10^6, 6\text{x}10^6]$ N/m, $K_2 = [5\text{x}10^6, 7.1\text{x}10^6]$ N/m, k = $8.8\text{x}10^6$ N/m. The interval width of the bearing properties is about 20%, a very conservative estimate. Indeed, for the rotor weight of 9000 N, two bearings are needed 60 mm diameter x 60 mm length. A H9/d9 fit, typical for this application, would yield interval midpoint radial clearance 85 µm and interval radius 35 µm. To achieve the assumed 10% interval radius of stiffness, a selection process during quality control would even be necessary to reduce the journal-bearing clearance interval radius.

The system is modeled with 3 lumped masses as shown, figure 2. The system equations are, with the nomenclature of figure 2,

$$m_1 x^{\circ\circ}_1 + (k/4 + K_1)x_1 - kx_2/2 + kx_3/4 = 0$$

$$m_2 x^{\circ}_2 - k^{\circ} x_1/2 + kx_2 - kx_3/2 = 0 \qquad (23)$$

$$m_3 x^{\circ\circ}_3 + kx_1/4 - kx_2/2 + (k/4 + K_2)x_3 = 0$$

Therefore, the system mass and stiffness matrices are

$$\mathbf{M} = \begin{bmatrix} m_1 & 0 & 0 \\ 0 & m_2 & 0 \\ 0 & 0 & m_3 \end{bmatrix},$$

$$\underline{\mathbf{K}} = \begin{bmatrix} k/4 + K_1 & -k/2 & k/4 \\ -k/2 & k & -k/2 \\ k/4 & -k/2 & k/4 + K_2 \end{bmatrix}$$

Matrix $\underline{\mathbf{K}}$ is an interval matrix due to the bearing interval stiffnesses K_1 and K_2. This is a very usual situation in slider bearings where the change in the lubricant temperature might cause a very substantial change in the bearing stiffness. For example, for a light turbine oil SAE 5, change of temperature from 50 to 60 °C changes the oil viscosity and the Sommerfeld Number by 60% (Dimarogonas 1988). This results in change of the bearing stiffness, for a full circular bearing (Dimarogonas 1992), by 50%. Moreover, variations in bearing clearances have an additional substantial effect on the bearing stiffness (Dimarogonas & Paipetis 1983).

To test the results, a Monte Carlo method was developed to approach the exact interval solution by solving the general optimization problem presented in the last section. To this end, point matrices were constructed by random selection of points within the interval values of the interval parameters. The corresponding eigenvalue problem was solved and the maximum and minimum values of each eigenvalue were updated after each

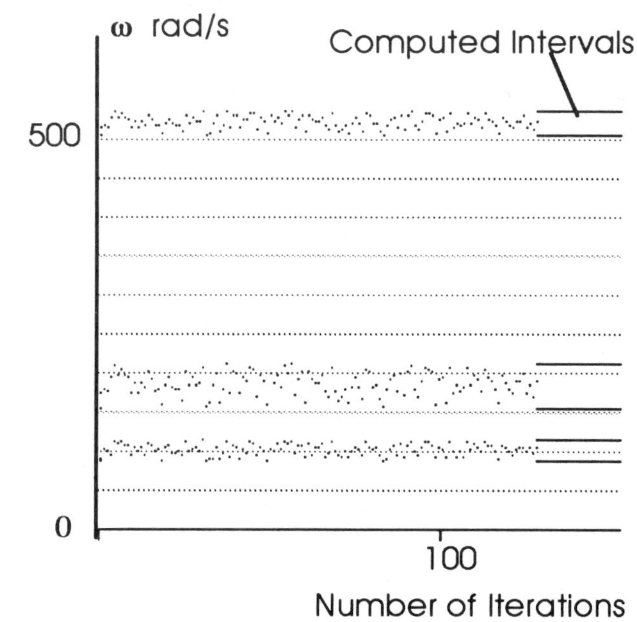

Figure 3: Monte Carlo and interval solutions of the interval eigenvalue problem for the rotor of figure 2.

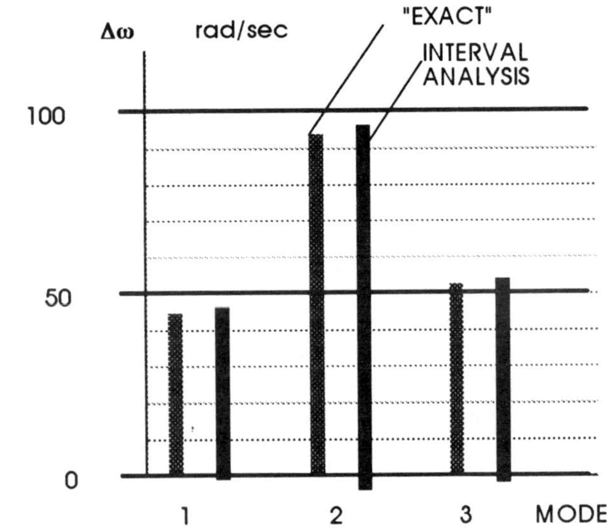

Figure 4. Comparison of the Monte Carlo solution and the results from equation 17 for the interval eigenvalue problem for the rotor of figure 2.

eigenvalue evaluation. The process converges rapidly, as shown in figure 3, where the interval eigenvalues ω are plotted versus iteration number. It can be noticed that the space between the first and second natural frequencies is almost diminished because of the modestly interval bearing properties. The first bending mode (the 3rd natural frequency) is substantially highr because the rotor of the particular example was stiffer than the bearings. For stiffer bearings, the space between the rigid body modes and the bending mode is also diminished. A more accurate account for the difference between the interval solution for the eigenvalues and the Monte Carlo solution are shown in figure 4. It is apparent that very good agreement between the interval evaluation and the Monte Carlo solution is obtained, in spite of the fact that the sign of one of the elements of one eigenvector in the interval evaluation changed.

FORCED VIBRATION

In the case of a forced vibration problem where an analytical solution exists, it was demonstrated above that an interval solution can be obtained by direct interval evaluation of the response function. For general excitation, where numerical methods have to be used, numerical integration results in a monotonously increasing solution interval and cannot yield usable results. Modal analysis, however, could be used based on the solution of the interval eigenvalue problem presented in the previous section. The modal matrix decouples the system interval equations and the solution of the interval modal equations is

$$Q_i(t) = E^{-Z_i\Omega_i t}\{Q_i(0)\mathrm{COS}\Omega_{di}t + $$

$$+[2Z_i\Omega_i Q_i(0) + Q^\circ_i(0)/\Omega_i]\mathrm{SIN}\Omega_{di}t\}$$

$$+\int_0^t W_i(s)G_i(t-s)ds \qquad (24)$$

were $\Omega_{di} = \Omega(1-Z_i^2)^{1/2}$,

$$G_i(t) = (1/\Omega_{di})e^{-Z_i\Omega_i t}\,\mathrm{SIN}\Omega_{di}t,$$

$$W_i(t) = \boldsymbol{\phi}_i^T\mathbf{f}(t)$$

$$Q_i(0) = \boldsymbol{\phi}^T\mathbf{Mx}(0)$$

$$Q_i^\circ(0) = \boldsymbol{\phi}^T\mathbf{Mx}^\circ(0)$$

After one period of oscillation, harmonic functions reach their bounds and then

$$Q_i(t) = UE^{-Z_i\Omega_i t}\{Q_i(0) + [2Z_i\Omega_i Q_i(0) + Q^\circ_i(0)/\Omega_i]\} +$$

$$\int_0^t W_i(s)(1/\Omega_{di})e^{-Z_i\Omega_i(t-s)}ds \qquad (25)$$

were U = [-1, 1].

Numerical evaluation of the integrals in equations 24 and 25 is evidently convergent since the integrals do not involve distributive expressions and the integration step ds can be selected arbitrarily small to assure monotonocity.

For a numerical application, the rotor of the previous section was considered with a forcing function which is constant over one-half of one natural period of vibration and has magnitude 1000 N. The results are shown in figure 5.

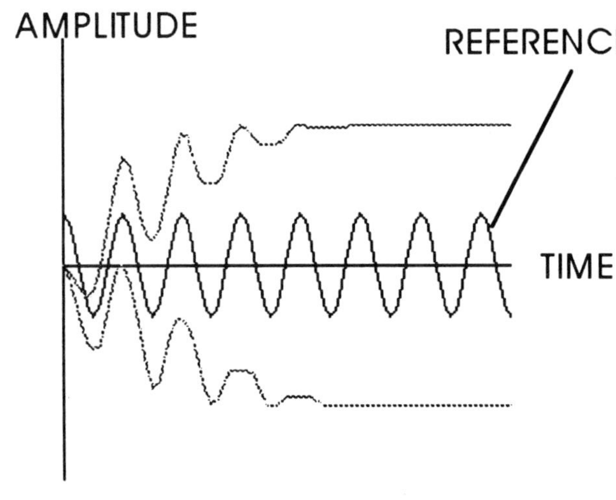

Figure 5: Interval forced response of the rotor of figure 2 to a finite impulse.

CONCLUSION

An interval systems vibration theory was presented. The natural and forced vibration problems for interval systems were discussed and solutions were developed using interval calculus.

It appears that analysis of interval systems involves merely the re-definition of the basic algebraic operations. However, the stability of the interval propagation due to the subdistributivity needs to be carefully examined in every problem, in particular when iterative methods are utilized.

A rational method for the solution of the eigenvalue problem in structural dynamics was presented. A Monte Carlo method was used to provide an exhaustive alternative to test the proximity of the evaluated interval solution to the true one. For small systems this is an attractive alternative, due to its simplicity, because interval calculus does not need to be used. For larger systems, application of the Monte Carlo method is computation intensive and its rate of convergence uncertain. For such systems, the optimization method presented here can be used. For systems with large interval widths, a converging iteration needs to be used. If the intervals are due to relatively small perturbations, the first iteration only is usually sufficient, since a convergence criterion was supplied, which can test whether the true value of the interval of the solution was reached during an iteration

The forced interval vibration problem was solved with modal analysis and it was shown that such solution is convergent.

REFERENCES

ALEFELD, G., HERZBERGER, J., *Introduction to Linear Computations*. New York: Academic Press, 1983.

APOSTOLATOS, N., KULISCH, U., Electron. Rechenanlagen 10, 73-78. Grunzüge einer Intervallrechnung für Matrizen und einige Anwendungen, 1968.

ARCHIMEDES (ca. 287-212 BC), *On the Measurement of the Circle*. In *The works of Archimedes*, edited by Sir Thomas L. Heath, Cambridge University Press, Cambridge 1897. Dover edition, 1953.

CHEN, J., *IEEE Transactions on Automatic Control*, 37, 4, pp. 541-544. Sufficient Conditions on Stability of Interval Matrices: Connections and New Results, 1992.

CHONDROS, T., DIMAROGONAS, A.D., *Journal of Vibration, Acoustics, Stress and Reliability in Design*, 111, pp 251-256. Dynamic Sensitivity of Structures to Cracks, 1989.

DEIF, S., *Z. angew. Math. Mech. 71, 1, pp 61-64*. The Interval Eigenvalue Problem, 1991.

DIMAROGONAS, A.D., PAIPETIS, S.A., *Analytical Methods in Rotor Dynamics*. Elsevier-Applied Science Publishers, London, 1983.

DIMAROGONAS, A.D., *Computer Aided Machine Design*. Englewood Cliffs, N.J.: Prentice Hall, 1988.

DIMAROGONAS, A.D., HADDAD, S., *Vibration for Engineers*. Englewood Cliffs, N.J.: Prentice Hall, 1992.

DWYER, P.S., *Linear Computations*. New York: J. Wiley and Sons, 1951.

HERZ, D., *IEEE Transactions on Automatic Control*, 37, 4, pp. 532-535. The Extreme Eigenvalues and Stability of Real Symmetric Matrices, 1992.

KULISCH, U., MIRANKER, W.L, *Computer Arithmetic in Theory and Practice*. New York: Academic Press, 1981.

MARKOV, S., ANGELOV, S., *Interval Mathematics, Int. Symposium, Freiburg, Germany. Berlin: Springer Verlag*. An Interval Method for Systems of ODE, 1985.

MOORE, R., *Interval Analysis*. Englewood Cliffs, N.J.: Prentice Hall, 1966.

NEUMAIER, A., *Interval Methods for Systems of Equations*. Cambridge: Cambridge University Press, 1990.

NICKEL, K., *Interval Mathematics, K. Nickel, Ed.. New York: Academic Press*. Bounding Eigenvectors of a Symmetric Matrix, 1980.

ROHN, J., *Z. angew. Math. Mech. 70, 6, pp 562-563*. Real Eigenvalues of an Interval Matrix with Rank One Radius, 1990.

SUNAGA, T., RAAG Memoirs II 2, 547-565. Tokyo: Gagujutsu Bunken Fukyu-kai. Theory of an interval algebra and its application to numerical analysis, 1958.

YOUNG,W.H., *Rendiconti Accademia di Lincei, Classe di Scienza Fiziche*, (5) 17, pp 582-587. Sull due funzioni a piu valori constituite dai limiti d' una funzione di variabile reale a destra ed a sinistra di ciascun punto, 1908.

YOUNG, R.C., *Mathematische Annalen* 104, 260-290. The Algebra of many-valued quantities, 1931.

DE-Vol. 63, Vibrations of Mechanical Systems
and the History of Mechanical Design
ASME 1993

STATIC AND DYNAMIC CHARACTERISTICS OF SPOT WELDED SHEET METAL BEAMS

Kaushik Pal
EASi Engineering
Bingham Farms, Michigan

Donald L. Cronin
Department of Mechanical and
Aerospace Engineering and
Engineering Mechanics
University of Missouri–Rolla
Rolla, Missouri

ABSTRACT

Although widely used as structural components in the auto industry, spot welded sheet metal beams manifest static and dynamic behavior that is not well characterized.

For the present study, sample beams of three representative cross–sections — hat, box and box with partition — were fabricated of sheet steel. The spacing between the spot welds that hold these sections together was varied from 25.4 mm (1 inch) to 203.2 mm (8 inch) using a 25.4 mm (1 inch) increment. The beams were subjected to static bending and static torsion tests, and bending and torsional stiffnesses were determined as functions of spot weld spacing. The beams were then vibrated, and significant lower natural frequencies were determined as functions of spot weld spacing. Mode shapes were also observed for these frequencies.

Pains were taken throughout the testing to ensure that the results obtained were of good statistical quality. Work was also done to distinguish the effect on results of beam length and end conditions from that of spot weld spacing.

As part of the study, finite element models of the beams were constructed. The results of finite element analysis (FEA) and experiment are compared and insights are offered concerning the appropriate modeling of such structures.

INTRODUCTION

In modern automobile body structures, a number of structural components are made by spot welding two or more pieces of sheet metal together. A typical resulting beam is illustrated in Fig. 1.

When a built–up beam of this type is deformed, the beam members are subjected to torsion and bending. In these cases, normal and shearing forces are transmitted through the spot welds.

Research on spot welded structures appears to focus either on the fatigue behavior or the static strength of spot welds. Fatigue behavior is described, for example, by Wilson and Fine (1981), Orts (1981) and Rossetto, *et al.* (1987).

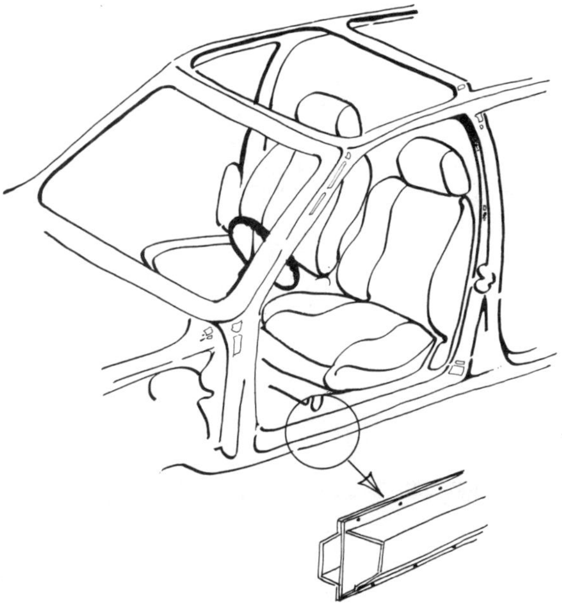

FIG. 1 A TYPICAL SPOT WELDED BEAM IN AN AUTOMOBILE

Jourmat and Roberts (1955), and Sawhill and Furr (1981) assessed the spot weld shear strength by static torsion tests. Ewing, *et al.* (1982) tested the static and impact strengths in bending of spot welded high strength low alloy (HSLA) and low carbon steel structures.

Mansoor (1988) and Tatsukawa, *et al.* (1989) evaluated the performance of spot welds in shear and normal stresses by non–destructive ultrasonic inspection.

Often quoted in the automotive industry is Myers (1966), who described static experiments to investigate the effect of spot weld spacing on torsional rigidity. He also investigated the effect on the bending rigidity of some typical automotive underbody cross members produced by introducing a non–uniformity in the cross section in the form of a "bump."

A more recent work by Lim, *et al.* (1990), reported on an analytic investigation of the torsional rigidities of thin–walled beams with spot welded joints. In their work, Lim, *et al.* modeled the spot welds as springs. They also showed that introducing a longitudinal partition between two "hat sections" increases bending strength, but tends to increase the stress in the spot welds and to reduce fatigue strength. Lim, *et al.* (1988), also observed that the partition decreased the torsional rigidity under the influence of a static moment.

Prior to the present study, no work appears to have been done to examine the effect on the dynamic response of spot weld spacing for such structures. In the study described here, a detailed statistical analysis of the data was carried out before conclusions were drawn. A wide range of spot weld spacings, 203.2 mm (8 inch) to 25.4 mm (1 inch), was examined. Prior to the present study, there also appears to have been no work done concerning changes in mode shapes due to changes in spot weld spacing.

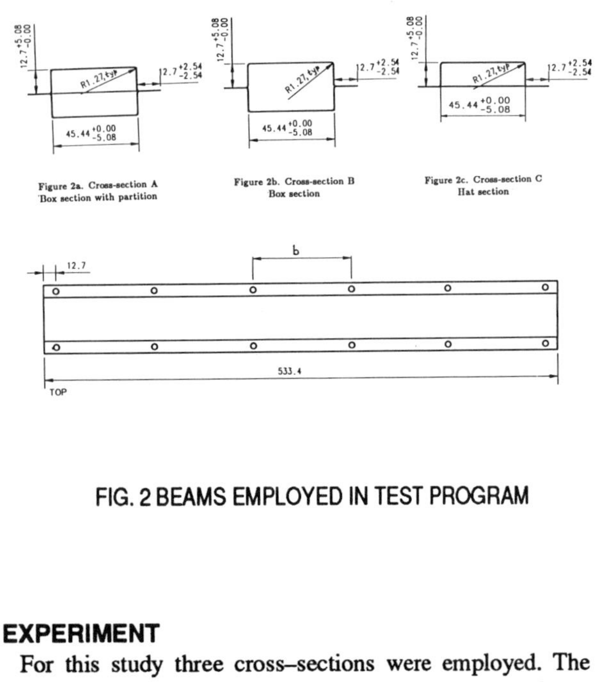

FIG. 2 BEAMS EMPLOYED IN TEST PROGRAM

EXPERIMENT

For this study three cross–sections were employed. The box cross–section with a partition shown in Fig. 2(a) will be referred to in the following as cross section A, that of Fig. 2(b), a box, as cross section B, and that of Fig. 2(c), a hat, as cross section C.

Five specimens of each cross–section were constructed and tested. The material employed was low carbon, cold rolled, matte finish, oiled 20 gage sheet steel, a typical automotive material. Twenty gage sheet has a thickness of 0.91 mm (0.036 inch). An additional five specimens of cross–section B were constructed which were 762 mm (30 inch) long. These specimens were used to investigate the effect of the change of length of the specimens on their dynamic behavior.

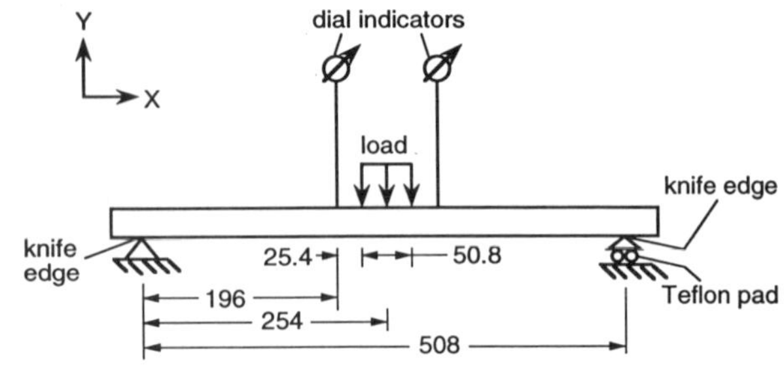

FIG. 3 SCHEMATIC OF BENDING TEST

STATIC TESTS

Bending tests

For the bending tests the specimens were arranged as simply supported beams as shown in Fig. 3.

A piece of Teflon was introduced between the ground and one of the supports to simulate a roller. The span of the beam between supports was 508 mm (20 inches). The load was uniformly distributed over the central 50.8 mm (2 inches) of the beam. The load was varied from 9.05 kg (20 lbs) to 72.4 kg (160 lbs) in increments of 9.05 kg (20 lbs) to obtain a force–deflection curve.

For the beams of cross–section C, the experiment was performed twice. In one case, the load was applied on the hat side of the beam, and in the other case the beam was flipped over, and the load was applied on the flat side of the beam.

A finite element model of the beam was used to verify that all loadings produced stresses in the elastic range.

Deflections were measured using dial indicators positioned 25.4 mm (1 inch) away from the load on both sides of the load, and averaged.

Torsion tests.

Fixtures were designed to mount the specimens on a Tinius Olsen LO–TORQ torsion test machine. The fixtures were designed so that the effect of local deflections on the results were minimized.

One end of a specimen was held fixed, while a torque was applied to the other end. The torque was varied from 1.127 Nm (10 in–lb) to 9.0 Nm (80 in–lb) in increments of 1.127 Nm (10 in–lb). A finite element model was used when designing the test to ensure that stresses would remain in the elastic range during the entire test.

Dial indicators were positioned on arms attached to each end of the specimens to obtain an accurate value for the angle of twist.

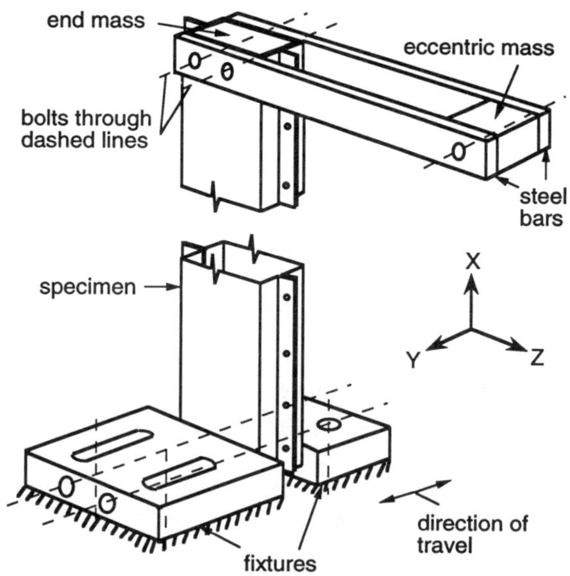

FIG. 4 SPECIMEN AND MOUNTING FOR DYNAMIC TEST

DYNAMIC TESTS

Mounting.

Each beam was attached as a vertical cantilever to a slip table driven by an electromagnetic vibration exciter. The experimental set–up is explained as follows:

The specimens were attached to the slip table using a fixture as shown in Fig. 4. Two steel bars and an eccentric mass were used to load the beam inertially. The eccentric mass used was 0.25 kg (0.555 lb). The distance that the eccentric mass was offset from the beam centerline was 111 mm (4.375 inch).

Structure was introduced at each end of the beams to distribute loading and to minimize local deformations of the sheet metal. A typical structure is shown in Fig. 4.

All bolts associated with the fixture were tightened with a torque wrench to help ensure that the tests were repeatable and reproducible.

Excitation.

An MB Dynamics shaker (Model PM250) with a TEAM (Model Number 3279204) slip table was used to excite the specimens. For the determination of natural frequencies a narrowband random vibratory signal generated by a Hewlett Packard 3582A spectrum analyzer was applied to the shaker via shaker electronics as the forcing signal. A sinusoid was used as the input signal for the mode shape determination.

Instrumentation.

To find the natural frequencies, one of two accelorometers was mounted on the eccentric mass. The other accelerometer was mounted on the base of the specimen. The spectrum analyzer was used to process the accelerometer signals to produce a transfer function for the beam. A plot of the transfer function showed peaks at the locations of the natural frequencies. These regions were zoomed using the analyzer and values of the first several natural frequencies noted. These values are assumed to be accurate to ± 0.04Hz.

To measure the mode shapes at a particular natural frequency, the frequency of the excitation was held constant, and an accelerometer was indexed along the length and across the width of the specimen. The displacement was thereby determined on a one inch (25.4 mm) grid with an MBIS (Model 2290) vibration monitor. A plot of the displacements gave a picture of the mode shapes.

DEVELOPMENT OF THE NASTRAN MODEL

Finite element models of the beams were developed using MSC/NASTRAN. Two types of plate elements, CQUADR's and CQUAD4's were evaluated. The CQUADR element, introduced in MSC Version 67, produced results within two percent of results produced by the CQUAD4 element. The performance of these elements was determined by running the models for a static deflection case. The CQUADR elements were selected for our analysis.

Several methods were examined to model the spot welds. The use of rigid bars was the first to be eliminated since it does not take into account the elastic behavior of the spot welds. Elastic rods which can deform in tension, compression and torsion were tried. These produced results that were significantly in error in a comparison involving a static deformation case. The spot welds were then modeled using a six–sided elastic solid element (CHEXA) with eight grid points. This model was acceptable as the results were comparable to within 10% of the experimental results involving a static deformation case.

There were about 12,000 degrees of freedom for beams of cross sections B and C. The corresponding degrees of freedom for the beam of cross section A was about 18,000. This mesh produced a maximum aspect ratio of 1.75.

For the static bending case, the load was applied across the central 50.8 mm (2 inches) of the beam. The load was uniformly distributed over 36 grid points using static load vectors. The deflections were measured 25.4 mm (1 inch) away from the load to avoid localized behavior and misleading results.

For the static torsion case, a massless rigid body was attached to the nodes at the free end of the cantilever beam. A moment was applied to the beam through the rigid body.

The beams were restrained in the following manner: When setting up the beam as a simply supported beam, four degrees of freedom were constrained to be zero for each of the nodes on both ends of the

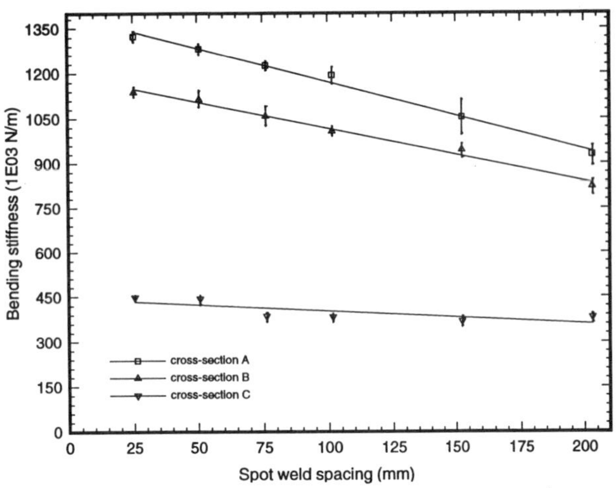

FIG. 5 BENDING STIFFNESS VERSUS SPOT WELD SPACING

beam that are in contact with the knife edge. The two degrees of freedom that were not constrained to be zero were the translational degree of freedom in the *X* direction (Fig. 3) and the rotational degree of freedom in the Z direction. When setting up the beam for the torsion test, all of the degrees of freedom for all of the nodes on the fixed end of the beam were constrained as zero. A rigid body was introduced at the other end and the torque was applied on it.

For the dynamic case, the cantilever end conditions described above were used. An eccentric concentrated mass was attached at the free end, offset by 111 mm (4.375 inch) using elastic beams and a concentrated mass. The concentrated mass assumption introduced an error since it did not take into account the moment of inertia of the eccentric mass about its own axis in the *X* direction (Fig. 4).

Dynamic analysis was performed using generalized dynamic reduction (GDR).

RESULTS AND DISCUSSION

In the following section are discussed the results obtained from the experiments and finite element analysis. Statistical analysis of the data as described in Mettlach (1991) and Snedecor and Cochran (1980) was carried out using the Student's *t* test. The Student's *t* test between non–independent paired measures gives an estimate of the difference in the paired measures, in the form of a confidence coefficient, which is explained later.

STATIC TESTS

Bending tests.

On conducting the experiments with the simply supported beams, we obtained sets of values of beam deflection for different values of applied load. The load was plotted against deflection for each of the specimens. A least squares best–fit line was used to estimate the slope of the load–deflection curve in each case, and these slopes were then averaged for five samples of each cross section (and orientation). Results are plotted versus spot weld spacing in Fig. 5. A least–squares straight line was used to fit the stiffness data shown. Higher order polynomials did not present any significant improvement of the fit.

Statistical analysis of the data for beams of cross section A and B indicate at least a 95% confidence coefficient in stating that beams of cross section A are stiffer in bending than beams of cross section B for a given spot weld spacing. Lim, *et al.* (1988) also observed that the partition increased the bending stiffness. A ninety five percent confidence coefficient means that 95 times out of 100, the statement above will be true. We can also conclude with at least an 85% confidence coefficient for beams of cross section A and B that the beams become stiffer in bending as the spot weld spacing is decreased.

Beams of cross–section A indicate a higher increase in bending stiffness as spot weld spacing is decreased than do beams of cross–section B. We argue this on the basis of confidence coefficients becoming larger as spot weld spacing decreases. This indicates a divergence of the populations of A and B bending stiffness for a given spacing.

An analysis of beams of cross–section C produced confidence coefficients of less than 50%. We conclude here that the bending stiffness of beams of this cross–section is relatively insensitive to changes in spot weld spacing. Similarly, data sets for the same beams loaded from the flat side or the hat side cannot be concluded to be different. The two data sets are thus combined and plotted. The standard deviations of beams of all three cross–sections are shown on Fig. 5 to give an idea of the spread of the data. Larger standard deviations were observed for spot weld spacings of 152.4 mm (6 inch) and 203.2 mm (8 inch) probably because when the spot weld spacing is not small with respect to the length of the beam, then end effects become relatively more significant. Those were also the first tests to be conducted and, for this reason, possibly more prone to experimental error.

Finite element analysis was used to model the static bending case. The results showed the same trends as shown by the experimental results. However, large deviations from the experimental results were observed at certain spot weld spacings. This is why the finite element results were not shown in Fig. 5. Both the finite element models and the experimental specimens exhibited buckling of the sheet metal between adjacent spot welds. Depending on the position at which the deflection was observed relative to the spot welds, this effect could contribute significantly. The problem with our finite element models is that buckling can occur such that one piece of sheet metal penetrates the other — this can not, of course, happen in reality. The result is a larger than realistic observed displacement and a consequent lower than realistic estimate for stiffness.

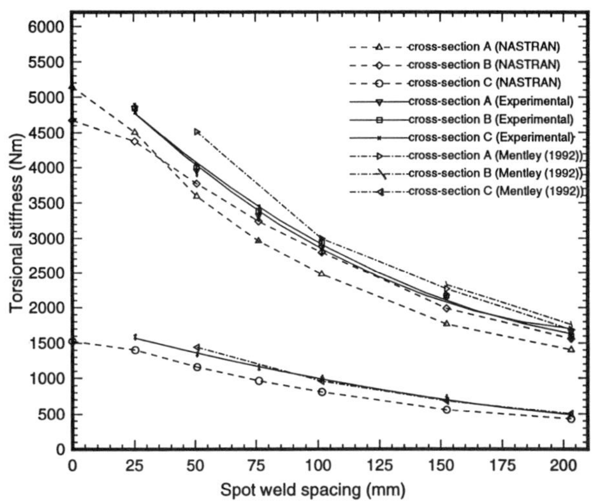

FIG. 6 TORSIONAL STIFFNESS VERSUS SPOT WELD SPACING

Torsion tests

On conducting torsion tests, we obtained sets of values of the angles of twist for different values of the torque. The applied torque was plotted against the angle of twist for each of the specimens. A least squares best–fit line was used to estimate the slope of the torque versus angle–of–twist curve for each beam. These slopes were then averaged for five samples of each cross section and are presented in Fig. 6. The results obtained from our finite element model are shown, and analytical results due to Mentley (1992) are also given.

The graphs clearly indicate that the torsional stiffnesses for all three cross–sections increase as the spot weld spacing is decreased. A best–fit polynomial of fourth order was used to fit the curves of the experimental values for all three cross sections. Lower order polynomials were evaluated and produced lower correlation coefficients. The standard deviations of beams of all three cross–sections are shown in Fig. 6 to give an idea of the quality of our results.

Statistical analysis of the data for all of the beams show at least a 99% confidence coefficient that all of the beams increase in torsional stiffness with decreasing spot weld spacing. However, beams of cross section B can be concluded to be more stiff than beams of cross section A with at least a 70% confidence coefficient for spot weld spacings between 50.8 mm (2 inch) and 101.6 mm (4 inch). For spot weld spacings outside of this range, the two sets of data can be concluded to belong to the same population.

Spot weld spacings of less than 101.6 mm (4 inches) happen to be the most commonly used range in the automotive industry. At the higher end of this range of spot weld spacing, beams of cross–section B can be concluded to be more stiff in torsion than beams of cross–section A. At the lower end, beams of cross–section A and B are found to have similar torsional stiffnesses. We can, thus, conclude that while the addition of the middle plate increases bending stiffness, it does not increase the torsional stiffness in many systems of practical interest. We can also conclude that the increase in torsional stiffness is higher than the increase in bending stiffness as the spot weld spacing is decreased from 203.2 mm (8 inch) to 25.4 mm (1 inch). Lim, *et al.* (1988) experimented with spot weld spacings between 40 mm and 60 mm and concluded that beams of cross–section B are stiffer than beams of cross–section A in torsion.

From Fig. 6, it is interesting to note that Mentley's (1992) analysis and our finite element analysis bracket the experimental results.

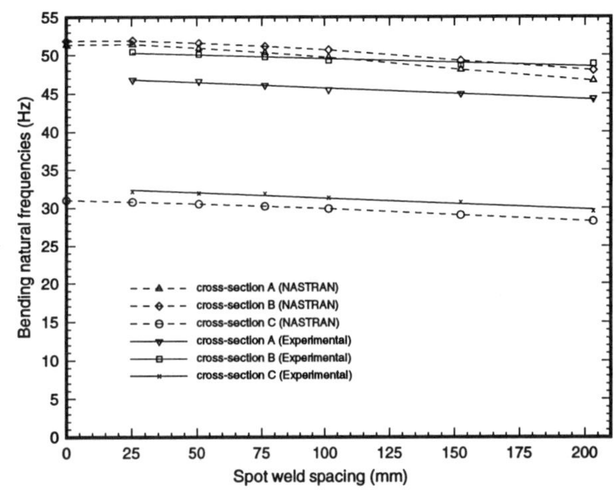

FIG. 7 BENDING NATURAL FREQUENCIES VERSUS SPOT WELD SPACING

DYNAMIC TESTS

For the dynamic tests, plots of the finite element results indicated that the first natural frequency for all of the beams has a corresponding mode shape that is primarily a bending mode. This bending occurs in the longitudinal direction (*Y* direction in Fig. 4) with respect to the beam. This is also the direction in which the beams are vibrated. These frequencies are referred to here as bending natural frequencies and are plotted against spot weld spacing (Fig. 7).

The second natural frequencies for cross sections A and B were also found to have corresponding mode shapes that are purely bending in nature. However, this bending occurs in the Z direction (Fig. 4), which is transverse to the beam. This direction is perpendicular to the direction of excitation of the shaker. Thus, this behavior was not observed during the experiment.

For cross sections A and B, the third natural frequency is coupled bending and torsion.

For cross section C, with spot weld spacings of 203.2 mm (8 inch) and 152.4 mm (6 inch), the second natural frequencies corresponded to coupled bending and torsion. For spot weld spacings less than 152.4 mm (6 inch), it was the third natural frequencies that corresponded to a coupled bending and torsion. These frequencies are referred to here as torsional natural frequencies and are plotted in Fig. 8.

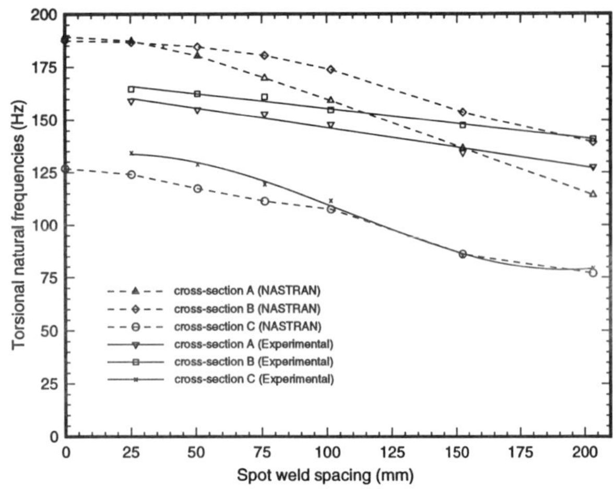

FIG. 8 TORSIONAL NATURAL FREQUENCIES VERSUS SPOT WELD SPACING

Statistical analysis of the data for all the beams show at least a 99% confidence coefficient that the beams of cross section B have a higher bending natural frequency than beams of cross section A. A 99% confidence coefficient was also observed in the bending natural frequency increasing with decreasing spot weld spacing for beams of cross–section A and C. For beams of cross–section B, confidence coefficients of 99% were observed only for spot weld spacings less than 152.4 mm (6 inch). Thus, we can conclude that in the range of spot weld spacing most common in the automotive industry, the bending natural frequencies increase with decreasing spot weld spacing.

A least–squares straight line was used to fit the data for all the beams for bending natural frequencies. Higher order polynomials did not present any significant improvement of the fit.

The plot of the torsional natural frequencies for beams of cross section A and B were again best–fit by a least–squares straight line, while those for beams of cross–section C were best–fit by a third order polynomial to account for changes in behavior with decreasing spot weld spacing.

The standard deviations of the experimental results of the dynamic tests were extremely small and are thus not shown on the plots. They suggest that the data is of very good quality.

A similar statistical analysis of the data for the torsional natural frequencies show at least a 99% confidence coefficient that the natural frequency increases with decreasing spot weld spacing for beams of all three cross–sections. The increase was found to be higher for beams of cross–section C than for beams of cross–section A and B as spot weld spacing decreased from 203.2 mm (8 inch) to 25.4 mm (1 inch).

A comparison between beams of cross section A and B show that for spot weld spacings higher than 50.8 mm (2 inch), the confidence coefficient is at least 99% for beams of cross section B having a higher torsional natural frequency than beams of cross section A. For spot weld spacings of 25.4 mm (1 inch), the confidence coefficient drops to 92% suggesting a possibility that the two curves are approaching, a behavior noted for the finite element analysis. Although we can not compare torsional stiffness to torsional natural frequencies, we do note a similarity in the trends of the plots.

There appears to be a larger discrepancy between the experimental observations and results predicted by the finite element analysis for the torsional mode than for the bending mode. From Fig. 8, curves for cross sections A and B tend to deviate from the finite element values as spot weld spacing is decreased. One possible explanation of this effect is that the influence of end conditions become more significant for higher frequency modes.

MODE SHAPES

The mode shapes were determined by observing the displacements at predetermined locations on the beam. No statistical work was done; only a single example of cross–section B was employed. Observations were made only for spot weld spacings of 203.2 mm (8 inch), 101.6 mm (4 inch) and 50.8 mm (2 inch) and for both the bending and the torsional mode shapes.

Mode shapes were compared using the modal assurance criterion (MAC)

$$MAC = \frac{(\langle\phi_1\rangle\{\phi_2\})^2}{(\langle\phi_1\rangle\{\phi_1\})(\langle\phi_2\rangle\{\phi_2\})}$$

where $\{\phi_1\}$ and $\{\phi_2\}$ are the two eigenvectors to be compared.

A MAC value of one indicates that the two eigenvectors are "parallel". A MAC value of zero indicates that they are orthogonal with respect to the identity matrix.

Bending Mode Shape.

Bending mode shapes for each of the above spot weld spacings were compared, and the corresponding MAC values are given in Table 1. Each of these mode shapes are also compared with the mode shape for a continuously welded beam from the finite element model. A continuously welded beam is characterized in the table as having zero spot weld spacing.

TABLE 1. BENDING MODE SHAPE MAC VALUES

Spot weld spacing (mm)	101.6	50.8	0
203.2	0.9976	0.9857	0.9910
101.6	1.000	0.9998	0.9878
50.8		1.000	0.9990

MAC values of two mode shapes can be found from the table by reading the row and the column of the two spot weld spacings to be compared. MAC values close to one indicate that no significant change in the mode shape has occurred with the change in spot weld spacing.

Torsional Mode Shape.

The third eigenvector couples bending and torsion. We refer to these eigenvectors as torsional eigenvectors. A comparison of the torsional mode shapes resulted in the MAC values given in Table 2.

TABLE 2. TORSION MODE SHAPE MAC VALUES

Spot weld spacing (mm)	101.6	50.8	0
203.2	0.9994	0.9875	0.9910
101.6	1.000	0.9925	0.9850
50.8		1.000	0.9570

MAC values of two mode shapes can again be found from the table by reading the row and the column of the two spot weld spacings to be compared.

EFFECT OF LENGTH OF THE SPECIMEN ON RESULTS

Beams of cross section B of length 762 mm (30 inch) were tested to determine whether or not our conclusions for the dynamic behavior of the beams are affected by the length of the beams. Five beams were tested in a manner similar to the shorter (21 inch) beams. This investigation was carried out for spot weld spacings of 203.2 mm (8 inch), 101.6 mm (4 inch) and 50.8 mm (2 inch) only.

No significant changes were observed in the bending natural frequency with changes in spot weld spacing. The torsional natural frequency was observed to increase weakly with a decrease in spot weld spacing. Finite element analysis was used to predict the behavior of these beams as well. The analysis predicted a stronger change in torsional natural frequency than we observed in our experiment, suggesting that clamped end conditions did not characterize our test well. No results produced by these tests affect our conclusions.

CONCLUSIONS

Based on our observations of spot welded beams, bending stiffness, torsional stiffness and the low–order natural frequencies associated with torsion all increase with decreasing spot weld spacing. Low–order natural frequencies associated with bending are relatively insensitive to spot weld spacing.

The bending stiffness of spot welded beams is a moderate to weak function of spot weld spacing (depending on the cross–section of the beam). Generally, the torsional stiffness of such beams has a relatively strong dependence on spot weld spacing. Low order natural frequencies associated with torsional behavior also have a relatively strong dependence on spot weld spacing.

The dependence of bending and torsional stiffness on spot weld spacing for our box beams and box beams with a partition was stronger than was the dependence of the same quantities on spot weld spacing for our hat section beams.

The use of a partition in a spot welded box beam increases the stiffness in bending about axes normal to the plane of the partition. The use of such a partition, however, degrades torsional stiffness for spot weld spacings typifying automotive applications (50 mm to 100 mm) and appears not to increase torsional stiffness for other spot weld spacings. Moreover, the partition lowers both low–order bending natural frequencies and torsional natural frequencies.

The influence of spot weld spacing on the mode shapes for spot welded beams appears to be very small. We believe that more work is in order here to incorporate a proper statistical look at the mode shapes.

Insofar as finite element modeling is concerned, we established that the use of solid elements to model spot welds produces the best comparison to experimental data for spot welded beams.

ACKNOWLEDGEMENTS

The authors wish to thank Mr. Norm Germeroth , Plant Manager of Detroit Tool Metal Products Company, Lebanon, Missouri, for his generous permission to use the company's equipment. We thank Mr. James Randolph and Mr. Mark Mathis of Detroit Tool for technical assistance. We wish to thank Mr. Thomas Cronin for helping generate the data sets needed for the finite element analysis, and also for lending a hand with the experiments. We also wish to thank Mr. Jeff Mentley of Rasna Corp. for supplying computed torsional data, and Mr. Mark Zebrowski of Ford Motor Company for discussions that led to the study.

REFERENCES

Ewing, K.–M. W., *et al.*, "Static and impact strengths of spot welded HSLA and low carbon steel joints," SAE paper 820281, presented at the International Congress and Exposition, Detroit, MI, Feb. 23–26, 1982.

Jourmat, P. and Roberts, J. E., "Assessment of spot weld quality by torsion test," *British Welding Journal*, Vol. 2, No. 5, 1955, pp 225–233.

Lim, *et al.*, "On stress distribution of spot welded box beams with a longitudinal partition subjected to torsion," *Int. J. of Materials and Product Technology*, Vol. 5, No. 1, 1990, pp 95–108.

Lim, *et al.*, "On torsional rigidity of spot welded box beams with a longitudinal partition," *Int. J. of Materials and Product Technology*, Vol. 3, No. 3/4, 1988, pp 359–369.

Mansoor, T. M., "Ultrasonic inspection of spot welds in thin gage steels," *Materials Evaluation*, Vol. 46, No. 5, April 1988, pp 650–658.

Mentley, J., Private Communications, September–November, 1992.

Mettlach, G. A., "Statistics of vehicle NVH response spectra from the statistics of tire and wheel non uniformities," Masters Thesis, University of Missouri – Rolla, 1991.

Myers, R. A., "Body Stress Analysis," SAE paper 660005, 1966.

Orts, D. H., "Fatigue strength of spot welded joints in a HSLA steel," SAE paper 810355, presented at the International Congress and Exposition, Detroit, MI, Feb. 23–27, 1981.

Rossetto, E., *et al.*, "The fatigue limit of spot welded lap joints," SAE paper 870640, presented at the International Congress and Exposition, Detroit, MI, Feb. 23–27, 1987.

Sawhill, J. M. and Furr, S. T., "Spot weldability tests for high–strength steel," SAE paper 810352, presented at the International Congress and Exposition, Detroit, MI, Feb. 23–27, 1981.

Snedecor, G. W. and Cochran W. G., *Statistical Methods.*, The Iowa State University Press, 7th Edition, 1980.

Tatsukawa, I., Satonaka, S. and Yamamoto, M., "Ultrasonic examination for soldered lap joints of thin sheet metals with focusing probe," *Yosetsu Gakkai Ronbunshu*, Vol. 7, No. 4, Nov 1989, pp 96–102.

Wilson, R. B. and Fine, T. E., "Fatigue behavior of spot welded high strength steel joints," SAE paper 810354, presented at the International Congress and Exposition, Detroit, MI, Feb. 23–27, 1981.

DE-Vol. 63, Vibrations of Mechanical Systems
and the History of Mechanical Design
ASME 1993

PASSIVE CONTROL OF VORTEX SHEDDING INDUCED VIBRATION OF A FIVE-HOLE PROBE

Neil B. Kimerer, Jr.
Applied Research Laboratory
Pennsylvania State University
State College, Pennsylvania

ABSTRACT

Vortex shedding in flow over a five-hole probe, used to measure three dimensional flow, excites the probe into vibration which causes measurement errors and acoustically induced structural fatigue of the probe stem. One method of reducing the coherent structure of the shed vortices and thereby reduce the vibration is to place a fin on the probe shaft with the fin free to rotate around the shaft. In a particular five-hole probe application there was no room for a fin so an alternate method of breaking up the coherent structure in the vortices was required. Coiling a wire around the probe was a technique tried using several wire diameters and coil spacings. Vibration levels were measured and analyzed for several wire coil configurations and compared to the vibration reduction achieved by the fin.

INTRODUCTION

Correlated vortex shedding from fluid flow around a circular cylinder creates unsteady forces on the cylinder inducing vibration, White (1979). This vibration can cause fatigue failure in the cylinder if the frequency of the vortex shedding is close to a resonance frequency. For a five hole probe, which is a cylindrically shaped instrument used to measure flow field velocities, calculations using methods found in Blevins (1986), indicated that correlated vortex shedding was going to generate undesired vibration in the probe. The normal solution for reducing the vibration levels of a five-hole probe is to install a fin on the probe shaft which breaks up the alternating forces around the probe. However, for this particular five-hole probe application, the fin could not be used because of the close proximity of the probe to rotating machinery. An alternate vortex dissipating technique was required. Walshe and Woolton (1970) presented a vortex dissipating technique consisting of a helical strake on a smoke stack. A modification of this technique was adapted for the five-hole probe case by wrapping a coil of wire around the probe stem. To determine the effectiveness of the coils to reduce the probe vibrations caused by the vortex shedding the probe was tested in a six-inch diameter water tunnel. Vibration levels of the bare probe were measured and used as a reference for the reduction in vibration due to treatments of the probe. Vibration levels for two fins and 12 wire coils were then measured. The attenuations of the vibration that resulted from the application of the fins and the coils are presented and discussed. Since the interest of this study was to compare vibration reduction, absolute vibration levels were not required.

BACKGROUND

A five-hole probe is an instrument used to determine the three dimensional flow vector of a point in a flow field. The probe is inserted into the flow field and pressure measurements are taken at the five ports in the probe tip. Because of the Bernoulli effect from the fluid flowing around the probe tip the five pressures are different for different flow velocities and angles, therefore, from these five pressures the flow field velocity vector can be determined. The five-hole probe was to be manufactured and used in accordance with the procedures in Treaster (1986), Welz (1986), and Treaster (1978). The cylindrical stem supporting the probe tip is excited into vibration by the correlated vortex shedding. The vibrations displace the measuring tip of the five-hole probe in the flow field. This displacement introduces errors in the measurements as well as the possibility of causing fatigue failure if the probe is excited near one of its natural frequencies.

The Strouhal number is a dimensionless parameter defined as

$$Strouhal\ No = \frac{f\,D}{U}$$

where f is the vortex shedding frequency, D is the cylinder diameter, and U is the free stream velocity. Over the range of velocities of interest the Reynolds numbers ranged from 4.1 x 10^4 to 8.1 x 10^4. In this Reynolds number range the Strouhal number is 0.21, White (1986). Using this Strouhal number the calculated shedding frequencies ranged from 202 Hz at 20 feet per second to 404 Hz at 40 feet per second. These shedding frequencies were sufficiently close to the natural frequency of the five-hole probe to require some treatment to reduce the vibration response by reducing the magnitude of the vortex shedding forces.

Since there was no room for a fin on the probe in the intended application a wire coiled around the full length of the probe stem similar to the strake used by Walshe and Woolton (1970) was considered as an alternate vortex dissipating technique. Since no information was available on the effectiveness of this technique in water an experiment was designed to measure the effectiveness of the wire coils as well as determine if an optimum configuration could be found.

A five-hole probe 0.25 inches in diameter and 2.75 inches long was mounted in a six-inch diameter water tunnel. The bare probe was exposed to five velocities and the vibration levels were measured. A fin with a 0.50 inch cord that covered the full length of the probe was tested next. A similar fin with a 0.25 inch cord along with 12 wire coils of various wire diameters and coil spacings were also tested. The results of these measurements are presented and discussed.

INSTRUMENTATION

To measure the effectiveness of the various vibration reducing hardware, a piezoelectric accelerometer was mounted on the base of the 0.25 inch diameter, 2.75 inch-long probe stem.The vibration levels were measured using the piezoelectric accelerometer with a charge amplifier connected to an SD 375 Spectral Dynamics analyzer. Photographs of the fins and coils can be seen in Figure 1.

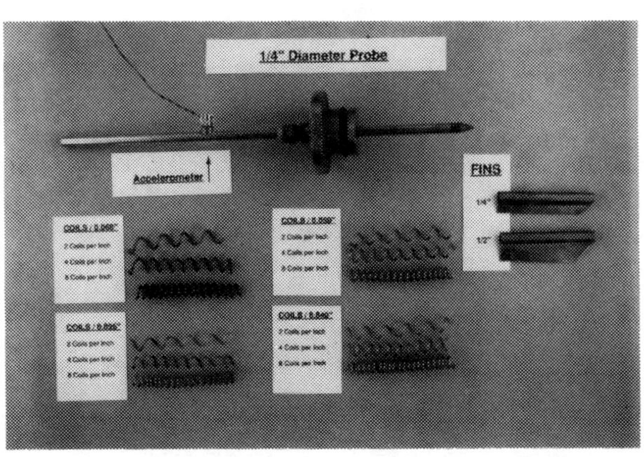

FIGURE 1

DATA ANALYSIS

Vibration levels at only two frequencies in the measured spectrum from 0 to 2000 Hz were above the background noise. A 245 Hz frequency was approximately 30 dB above the background noise and a 1390 Hz signal was about 25 dB above background noise. A sample of the accelerometer spectrum from the bare probe flow tests is plotted in Figure 2. This spectrum is the average of 100 samples. All the data in Figures 3 through 8 are corrected for reference to one volt. These two frequencies were the only two frequencies considered in the following analysis. It is recognized that the resonate frequencies of the five-hole probe system were affected by the added mass of the accelerometer, fins and coils as well as the change in stiffness resulting from the fins and coils. However, since the main interest of this study was to determine the reduction in five-hole probe system resonate vibration amplitude resulting from the application of the fins and coils, the frequencies at which the five-hole probe system resonated were not of prime concern.

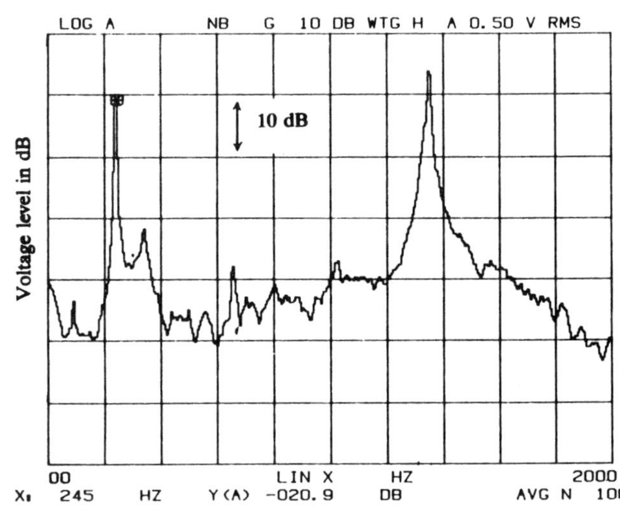

FIGURE 2

It can be seen from Figure 3 that the maximum vibration levels of the bare probe occurred at 25, 30, and 35 feet per second. Also from Figure 3 it can be seen that the vibration levels of the bare probe at 20 feet per second and 40 feet per second were lower than the maximum vibration levels by 4 dB and 15 dB respectively. This indicates that less attenuation at the 20 and 40 feet per second velocities is required to keep the vibration levels low enough to avoid fatigue.

The vibration levels of the five-hole probe with the coils and the however, their studies were all done in air where cavitation was not a problem.

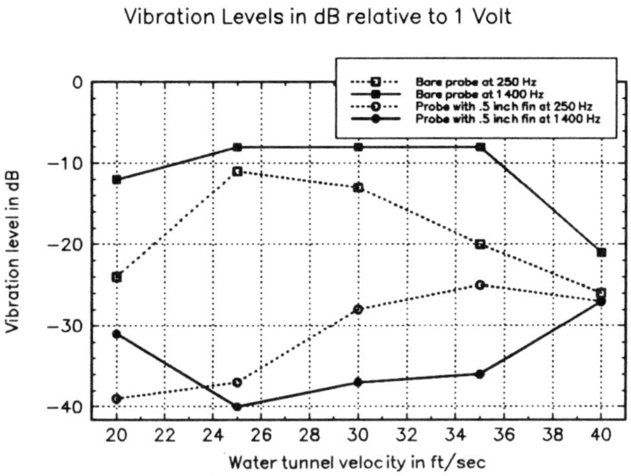

FIGURE 3

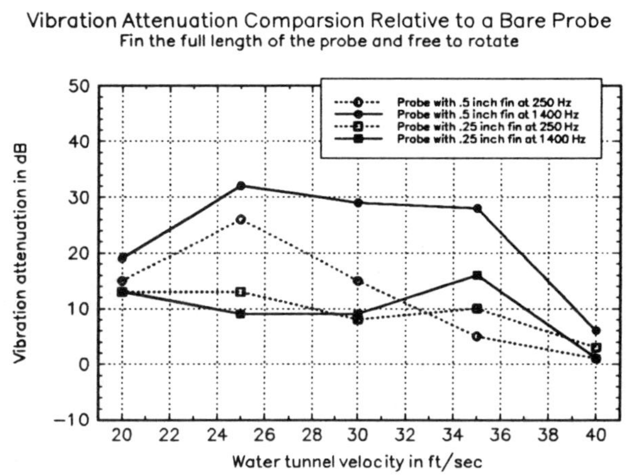

FIGURE 4

fins installed were compared to the vibration level of the bare probe and the results can be seen in Figures 4 through 8. In these figures the attenuations resulting from the fin or coil are plotted versus flow. In Figure 4 it can be seen that the 1400 Hz vibration for the 0.50 inch cord fin is reduced by about 30 dB at the 25, 30 and 35 feet per second velocities. Note that the attenuations given in Figure 4 are the difference between the bare probe vibration levels and the 0.50 inch cord fin vibration levels found in Figure 3.

The 0.50 inch cord fin reduced the vibration of the five-hole probe at 20 and 40 feet per second by about 20 dB and 4 dB respectively. From Figure 4, attenuation at 250 Hz was about 26 dB at 25 feet per second but diminished to 2 dB at 40 feet per second. The 0.25 inch cord fin never achieved an attenuation above 14 dB at either frequency over the velocity range except at 35 feet per second where the attenuation of the 1400 Hz vibration was attenuated by about 16 dB. The overall effectiveness of the 0.25 inch cord fin was poor.

Similar comparisons were made for the five-hole probe with the coils installed. Two parameters were changed on the coils. Namely, the wire diameter and the coil spacing, or the helix angle. Four wire diameters of 0.035, 0.040, 0.050, and 0.060 inches and coil spacings of two coils per inch, four coils per inch, and eight coils per inch were used. The 0.035 inch wire diameter is 14% of the probe diameter. This is close to the 10% ratio of the strake width to the cylinder diameter used by Walshe and Woolton (1970) on the chimney strake. Walshe and Woolton noted that rounded edges of the strakes produced less correlated vortex shedding vibration reduction than sharp edged strakes, however, their studies were all done in air where cavitation was not a problem.

Walshe and Woolton did mention the use of spiralled cables and noted that they were less effective than the sharp edged strakes. Wire was selected for the strake in our experiments because it was readily available, easy to shape, and the circular cross section would minimize cavitation.

In Figure 5 the attenuation for the three coils spacings for the 0.035 inch diameter wire are plotted for both frequencies.

At the 20 feet per second velocity, all three coil spacings produced significant attenuation, however at the three velocities where the vibration levels were the greatest the attenuation was poor. An attenuation of 24 dB of the five-hole probe vibration was achieved by the 0.035 inch diameter wire for the 250 Hz frequency at 25 feet per second but this improvement in vibration reduction rapidly fell to about 4 dB by 30 feet per second. The overall attenuation in the 25 to 35 feet per second range was below 8 dB. Similar results were obtained with a 0.040 inch diameter wire.

Figure 6 presents the data for the 0.050 inch diameter wire. Again, attenuation of the vibration levels at the 20 feet per second velocity are good. However, with this wire diameter the coil spacings begin to show differences in the attenuation particularly at the higher flow velocities. Similar results were obtained with a 0.060 inch diameter wire.

At the 1400 Hz frequency the two coils per inch spacing showed higher attenuations at the 20, 25, 35, and 40 feet per second velocities. The only attenuation that was higher than the two coils per inch was the eight coils per inch at 30 feet per second. For the 250 Hz frequency the two coils per inch vibration attenuation was equal to or slightly better than the two

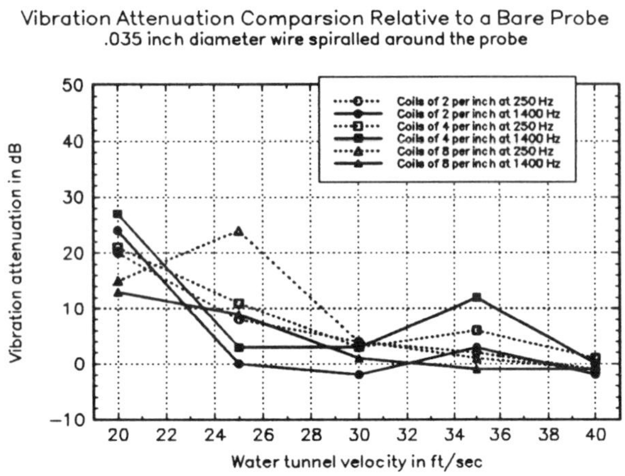

FIGURE 5

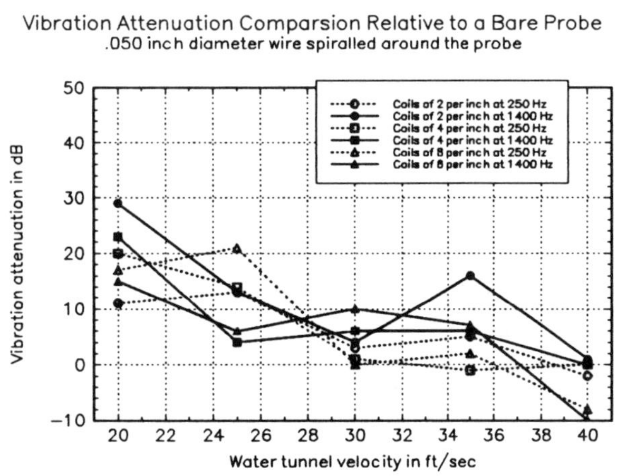

FIGURE 6

other coil spacings except at 25 feet per second where the eight coils per inch was better. At 40 feet per second the attenuation for all three coil spacings was about zero.

Attenuation values for the four wire diameters with the 2 coils per inch spacing at the 250 Hz frequency are plotted in Figure 7. Attenuation for the two fins are also plotted in Figure 7. From Figure 7 it can be seen that the 0.060 inch wire gave the best attenuation at all velocities with the exception of the 0.50 inch cord fin. At 35 feet per second the two coils per inch strake gave about the same probe vibration attenuation as the 0.50 inch cord fin. Though the 0.25 inch cord fin at the 35 foot per second velocity proved to be slightly better than either the 0.50 inch cord fin or the two coils per inch strake this was the only condition where the 0.25 inch cord fin showed significant attenuation. Since neither fin can be used because of the space considerations the 0.060 inch diameter wire at 2 coils per inch was the best performing configuration at 250 Hz.

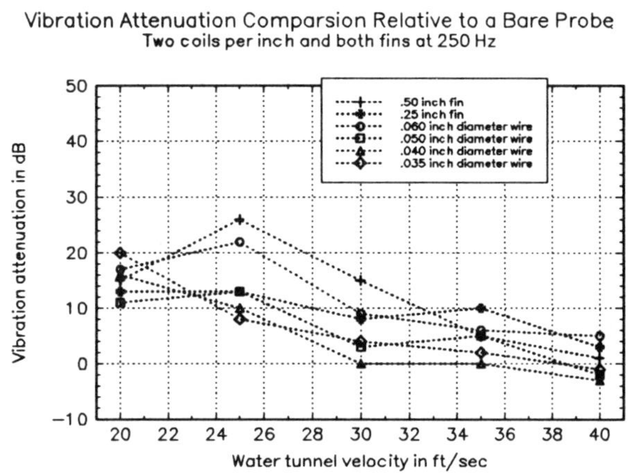

FIGURE 7

Plots similar to those in Figure 7 can be seen in Figure 8 for the 1400 Hz frequency. Here the 0.50 inch cord fin showed the highest attenuation. Though these data show mixed results at different velocities, the 0.060 inch diameter wire showed about the same attenuation as the 0.050 inch diameter wire and the 0.25 inch cord fin. The smaller wire diameters showed less attenuation than the 0.050 and the 0.060 inch diameter wires.

CAVITATION

Incipient cavitation occurred at 35 feet per second and 20.4 psia on all the configurations presented. Cavitation was suppressed by increasing the water tunnel pressure for the 0.060 inch diameter wire at two coils per inch and 40 feet per second velocity. Desinent cavitation disappeared at 40 feet per second and 28.0 psia on the five-hole probe with the 0.060 inch wire strake with two coils per inch. The suppression of the cavitation showed no shift in frequency in either of the two frequencies considered and no change in the amplitude of the 250 Hz frequency. The 1400

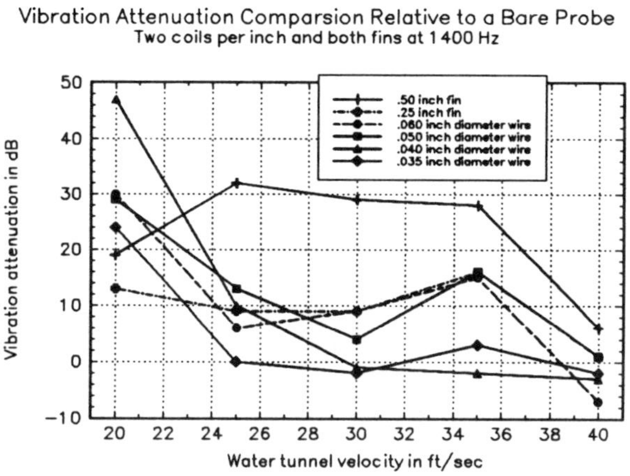

FIGURE 8

Hz frequency increased in amplitude about 5 dB with the suppression of the cavitation at the 40 feet per second water velocity.

CONCLUSION

Significant reductions in the vibration of five-hole probes excited by vortex shedding were achieved with the wire coils, however, the vibration attenuation was not as great as that achieved with the 0.50 inch cord fin. Though the 0.50 inch cord fin gave the most attenuation over the whole range of velocities and frequencies it could not be used because of the probes close proximity to rotating machinery. Though the vibration attenuations with the coils were not uniform for all flow and frequencies they were sufficient to prevent fatigue of the probe stem. The strake made from the 0.060 inch diameter wire at two coils per inch gave the best overall performance and was the configuration selected for the test application.

REFERENCES

White, F. M., 1979, "Fluid Mechanics," McGraw-Hill Book Company, (New York, New York) pp. 278-279

Blevins, R. D., 1986, "Flow-induced Vibration," Robert E. Krieger Publishing Company (Malabar, Florida) pp. 11-54

Walshe, D. E., and R. L. Woolton, 1970, "Preventing Wind Oscillations of Structures of Circular Section," Proc. Inst. Civil Engrs. (London)

Treaster, A. L. and Yocum, "The calibration and Application of Five-hole Probes", ISA Transactions, Vol 18, No 3, pp23-24, May 1979

Welz, J. P., "An Algorithm for Using the Five-hole Probe in the Nonnulling Mode", Proceedings of the AIAA/ASME Fluid Measurement and Instrumentation Forum, FED-Vol 34, May 1986

Treaster, A. L. and H. E. Houtz, "Fabricating and Calibrating Five-hole Probes", Proceedings of the AIAA/ASME Fluid Measurement and Instrumentation Forum, FED-Vol 34, May 1986

DE-Vol. 63, Vibrations of Mechanical Systems
and the History of Mechanical Design
ASME 1993

DETERMINATION OF THE POSITION OF SOLID/LIQUID INTERFACE DURING SOLIDIFICATION AND MELTING OF METALS WITH VIBRATIONAL PARAMETERS

Wang Fengquan, Zhu Zhenghua, and Chen Shiyu
Department of Mathematics and Mechanics
Southeast University
Nanjing, China

ABSTRACT

The paper proposes the method to locate the solid / liquid interface with vibrational parameters during solidification for the first time. The sufficient difference in resistance to shear stresses between liquid and solid phases of metals and alloys permits the application of vibrational parameters to locate the interface in real time and in situation during solidification. Based on the solidification theory, continuum mechanics, vibrational modal analysis and sensitivity analysis, the author established the mechanical model and the dynamical equations of typical Bridgman solidifying system, derived the sensitivity of eigenvalues of the Bridgman system to the location of the solid / liquid interface as well as the calculating formulae concerned. The experimental results are quite accordant with those of computed ones.

INTRODUCTION

The solid / ligiud interface during solidification is the isothermal plane temperature of which equals to melting point of the solidifying material. From the point of view of macro mechanical properties, the solid / liquid interface can be defined as the interface between solid and liquid phases of solidifying material. Since the control of solidification depends on the control of solidifying parameters, such as the solid / liquid interface moving velocity, the temperature gradiant, it is of great importance both in research and in production of solidification to locate the solid / liquid interface in situation and in real time.

At the pressent time, the method used to determine the position of solid / liquid interface is metallography. It does not give satisfaction to the requirement of real–time analysis and control in the field, for it relys on a postmorten autopsy. Starting with around 1965, ultrasonic pulse–echo techniques have been used to locate the solid / liquid interface during solidification and melting of metals [1]. These techniques are based on the acoustic impedance difference between solid and liquid phases. It is known that only the solid phase possesses shear strength. Hence, the dynamic properties of solidifying system (include: eigenvalues, eigenvectors) vary with the solid / liquid interface position during solidification. The sufficient difference in resistance to shear stress between solid and liquid phases of metals permits the application of vibrational parameters to determine the position of solid / liquid interface in real time and in situation.

BUILD MODEL

The Bridgman solidifying system used here consists of a crucible [2], a crystallizer and a specimen as shown structurally in Fig.1(a). The bottom of the crucible which contains the specimen is tightly connected with the crystallizer. Such system can be simplified into a model of cylindrical shell with variable cross section and partly filled with fluid as shown in Fig.1(b). Compared with the general

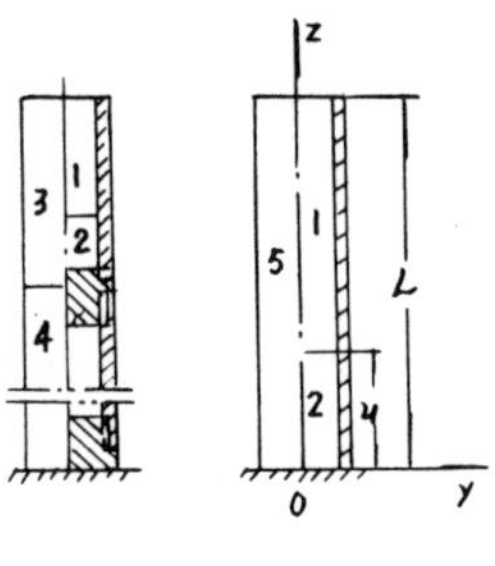

Figure 1 Bridgman Solidifying System
1 liguid, 2 solid, 3 crucible, 4 crystallizer, 5 cylindrical

model of shell containing fluid, it has two distinguishing features: The first, part of the model, esp. the solid / liquid interface is always in a closed opaque high temperature field. The second, the model is a time variant system, for its stiffness varies with the ratio of solid to liquid.

In order to locate the solid / liquid interface during solidification with vibrational parameters effectively, we analyse the transverse vibration of the model. under the axisymmetric load, the fundamental equation of cylindrical shell is reduced greatly. The transverse vibration equation is (1), where Φ_r is the radial displacement, $a$is the shell radius, δ is the shell thickness, ν is the poission ratio, P_r is the radial external load, E is the Yong's modulus.

$$R\frac{\partial^4\Phi_r}{\partial z^4}+\frac{E\delta}{a^2}\Phi_r=P_r,\quad R=\frac{E\delta^3}{12(1-\nu^2)} \tag{1}$$

The molten metal, i.e, the interior fluid is considered as ideal and incompressive. The crucible in Bridgman method does not revolve so that the forced convection can be neglected. In addition, the direction of temperature gradient is opposite to that of gravity and the density of molten metal in the lower part of crucible is heavier than that of in the upper part, so the natural convection is taken no account either. According to previous study on coupled vibration of cylindrical shell filled with fluid [3], the effect of the sway of moltenmetal can be neglected. In consideration of all discussed above, the Bridgman solidifying system is regarded as a "liquid–solidified" elastic cylindrical shell. Its transverse vibration can be dealed with the transeverse vibration of non–uniform beam. The centrifugal force, the Coriolis force, the axial force caused by temperature and internal pressure, the effect of shear force, rotating force and inertia and damping are all neglected. The transverse free vibration equation of the model is (2), where E is the Yong's modulus, I is the moment of inertia of the cross section, ρ is the mass density, A is the area of cross section, t is the time, $y(z,t)$ is the function of the trasverse vibration, subscripts w.s.L represent wall of shell, solid phase, and liquid phase respectively.

$$(E_W I_W+E_S I_S)\frac{\partial^4 y(z,t)}{\partial z^4}+(\rho_W A_W+\rho_S A_S+\rho_L A_L)\frac{\partial^2 y(z,t)}{\partial t^2}=0 \tag{2}$$

SENSITIVITY ANALYSIS OF MODAL PARAMETERS

During solidification, the variation of the length ratio of solid to liquid phases in crucible results in correspondent variation of the modal parameters of the Bridgman solidifying system, in Reference [4], we learn that the sensitivity of eigenvectors with respect to structural parameters can be expressed by the product of normal modes. Due to the limitation of Bridgman furnace, it is impossible to measure the normal modes. Therefore, the eigenvectors can not be adopted to locate the solid / liquid interface nor can modal mass, modal stiffness and modal damping. By contrast, it is possible to detect the natural frequency outside the furnace during solidification. The emphasis on analysis is given to the sensitivity analysis of eigenvalues, i.e., the relationship between natural frequency and the position of the solid / liquid interface. Perhaps the simplest approach to the solution of eigenvalues is to use Ritz's method. The boundary conditions are

$$y(o,t)=0,\ \partial y/\partial z|_{z=0}=0,\ \left.\frac{\partial^2 y}{\partial z^2}\right|_{z=L}=\left.\frac{\partial^3 y}{\partial z^3}\right|_{z=L}=0$$

Since we desire only the fundamental frequency, the function satisfies all boundary conditions.

$$Z=6L^2z^2-4Lz^3+z^4$$

$$U_{max}=\int_0^u(E_W I_W+E_S I_S)(Z'')^2dz+\int_u^L E_W I_W(Z'')^2dz$$
$$=\frac{9\pi E_W}{20}\{(D^4-d^4)L^5+\frac{E_S}{E_W}[d^4(L-u)^5-d^4L^5]\}$$

$$T_{max}=\int_0^u\rho_S A_S Z^2dz+\int_u^L\rho_L A_L Z^2dz+\int_0^L\rho_W A_W Z^2dz$$
$$=\frac{\pi(D^2-d^2)}{4}\rho_w(104/45)L^9+\frac{\pi d^2}{4}\rho_S$$
$$\cdot(7.2u^5L^4-8u^6L^3+4u^7L^2-u^8L$$
$$+u^9/9)+\frac{\pi d^2}{4}\rho_L[(104/45)-7.2u^5L^4$$
$$+8u^6L^3-4u^7L^2+u^8L-u^9/9]$$

Thus, we obtain the approximate fundamental frequency as

$$\omega^2(r)=\frac{1.8E_W D^2}{\rho_w L^4}\{[1-q^4(1-r)^5]\}/$$
$$\{(104/45)[1+(104/45)(\zeta-1)q^2(104/45$$
$$-7.2r^5+8r^6-4r^7+r^8-r^9/9)]\} \tag{3}$$

where ζ is the ratio of liquid density to crucible wall density, q is the ratio of the inner diameter (d) to the outer diameter (D) of the crucible, L is the total length, u is the length of solid phase, r is the ratio of u to L (include: crucible and crystallizer).

Taking the initial state of the solidifying system ($r=0$) as a base line, we get the dimensionless ratio of frequency.

$$\lambda=\lambda(r)=\omega(r)/\omega(0)=\{[1-q^4(1-r^5)][1+(\zeta-1)q^2]\}^{\frac{1}{2}}\cdot$$
$$\{(1-q^4)[1+45/104)(\zeta-1)q^2(104/45-7.2r^5+8r^6$$
$$-4r^7+r^8-r^9/9)]\}^{-\frac{1}{2}} \tag{4}$$

Let S denote the sensitivity of the ratio λ (i.e., frequency) to the ratio r (i.e., length of solid phase).

$$S=S(\omega/r)=\frac{\partial\lambda}{\partial r}=\frac{1}{\omega(0)}\frac{\partial\omega(r)}{\partial r} \tag{5}$$

$$\frac{\partial\omega(r)}{\partial r}=\frac{1.8E_W D^2}{\rho_w\cdot 2\omega L^4}\cdot\{[5q^4(1-r)^4][(104/45)+(\zeta-1)q^2$$
$$(104/45-7.2r^5+8r^6-4r^7+r^8-r^9/9)]-[1-$$
$$q^4(1-r)^5][q^2(\zeta-1)(-36r^4+48r^5$$

$$- 28r^6 + 8r^7 - r^8)] / [(104/45)(\zeta - 1)$$
$$q^2(104/45 - 7.2r^5 + 8r^6 - 4r^7 + r^8 - r^9/9)]^2\} \quad (6)$$
$$S = 0,$$
$$\partial\omega / \partial r = 0, \quad (7)$$

Thus

$$5q^2(1-r)^4[(104/45) + (\zeta - 1)q^2(104/45$$
$$- 7.2r^5 + 8r^6 - 4r^7 + r^8 - r^9/9]$$
$$- [1 - q^4(1-r)^5](\zeta - 1)(- 36r^4 + 48r^5$$
$$- 28r^6 + 8r^7 - r^8)] = 0 \quad (8)$$

Since

$$S|_{r=0} = \omega'(0) / \omega(0)$$
$$= \frac{1.8E_W \cdot (104/9) \cdot [D^2 + (\zeta - 1)d^2]q^4}{\rho_W \cdot 2\omega L^4[(104/45) + (\zeta - 1)q^2 \cdot (104/45)]^2} > 0,$$
$$S|_{r=1} = \omega'(1) / \omega(0)$$
$$= \frac{1.8E_W \cdot 9D^4(\zeta - 1)d^2}{2\rho_W \omega L^4 (104/45)^2} < 0,$$

furthermore, S is a continuous elementary function.

Thus, $\forall r \in [0,1]$, $\partial\lambda / \partial r = 0$, $\lambda|_{r=r_m} = \lambda_{max}$, $S|_{r=r_m} = 0$,

that is, at the standing point $r = r_m$, though λ reaches its extreme value, the vibrational parameters are useless in the identification of the solid / liquid interface. Since $\lambda = \lambda(r)$ is a higher–degree polynomial of r, λ_{max} is a global extreme value and the variation of function S is not small, we can apply vibrational parameters to locate the solid / liquid interface in spite of the existence of standing point. It is imperative to prove that $\lambda = \lambda(r)$ is a normal convex function within the range of r.

From the point of view of mathematics, it is a non–linear programming with restraint, i.e., under the restraint

$$g_j(x_1, x_2, \cdots, x_m) \geqslant 0 \quad (j = 1,2,\cdots,m)$$

For finding the optimum solution $(\overline{X}_1, \overline{X}_2, \cdots, \overline{X}_n)$, we make the non–linear objective function $f(X_1, X_2, \cdots, X_n)$ which is equal to its minium (or maximum) $f(\overline{X}_1, \overline{X}_2, \cdots, \overline{X}_n)$. Hence the result function is $\lambda = \lambda(r) = f_1(r) / f_2$, a fractional function.

TABLE 1 Numeracal results of Equation.4.

r	λ	r	λ	r	λ	r	λ
.00	1.000	.26	1.574	.52	1.645	.78	1.488
.02	1.087	.28	1.590	.54	1.640	.80	1.469
.04	1.162	.30	1.604	.56	1.634	.82	1.449
.06	1.227	.32	1.616	.58	1.626	.84	1.429
.08	1.284	.34	1.627	.60	1.617	.86	1.409
.10	1.334	.36	1.635	.62	1.607	.88	1.387
.12	1.378	.38	1.642	.64	1.596	.90	1.366
.14	1.417	.40	1.647	.66	1.584	.92	1.344
.16	1.452	.42	1.650	.68	1.570	.94	1.322
.18	1.483	.44	1.652	.70	1.555	.96	1.300
.20	1.510	.46	1.652	.72	1.540	.98	1.278
.22	1.534	.48	1.652	.74	1.523	1.00	1.255
.24	1.555	.50	1.649	.76	1.506		

According to the theorem of convex function and generzlized convexity, the convexity of $\lambda = \lambda(r)$ depends on its numerator $f_1(r)$, another linear fractional function, for its non–negative denominator f_2 is independent of r. The numerator $f_1(r)$ can be divided into numerator and denominator part further. Since the numerator part is a non–negative concave function, the denominator part is a positive convex function, thus, $f_1(r)$ is a pseudo–concave function and λ_{max} is a global extreme value. Let $q = 0.9, \zeta = 0.45$, $r = ih$, $i = 1,2,\cdots$, 100, $h = 0.01$. Find the solution of Equation.(8) with numerical method, such as Newton's method. we obtain $r_m = 0.4564$, $\lambda(0.4564) = \lambda_{max} = 1.653$. The computed results are shown in Table.1, and the correspondent curve of $\lambda(r)$ is shown in Fig.2.

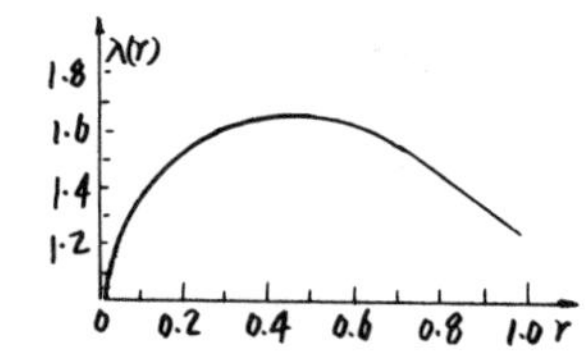

Figure 2 The curve of $\lambda(r)$

Fig.2 demonstrates that the sensitivity of natural frequency S is the slope of tangent of point on the curve. It can be noted that S is neither keeps constant nor changes monoly. In accordance with its magnitude, S can be classified into three categories: "sensitive", "blind" and "dull" during solidification. In application, S can be fallen into "sensitive" region by adjusting structural parameters, e.g. the ratio of the solid phase to the liquid phase.

EXPERIMENTAL PROCEDURES

Fig.3 shows structurally the furnace, the excitation and detection instruments for the tests made on Al–3 pct Mg alloy. The arrangement is basically a modified Bridgman crystal growth furnace design, which is chosen because it allows unidifiectional solidification and melting with virtually planar interface, controlled solidifying parameters and fast liquid quenching. The alundum crucible was heated by an electrical resistance furnace. During solidification, the position of the solid / liquid interface can be altered through adjustment of the level of cool water.

Several types of excitation were tested (random, pseudo random, etc.). Impulse tests were carried out with a dual channel signal processing system chiefly composed of an IBM–PC which has zoom function to obtain modal parameters. At the beginning of the solidification, while the level of cool water was at its lowest position, the Al–3 pct Mg alloy specimen was heated to melt partially and a steady thermal field was eatablished both in furnace and in crucible.

Then, the solidifying system was excited by hammer impulse at the point on the attainable section of crystallizer, simultaneously the response of the system was picked up by the accelerometer. The sampling time is 1 ms. Both force and response signals were sent into processing system and analyzed. Fig.4 shows the magnitude of the frequency response function in coincidence with a given position of the solid / liquid interface. Modal parameters can be identified through frequency response function. In this paper, eigenvalues are identified.

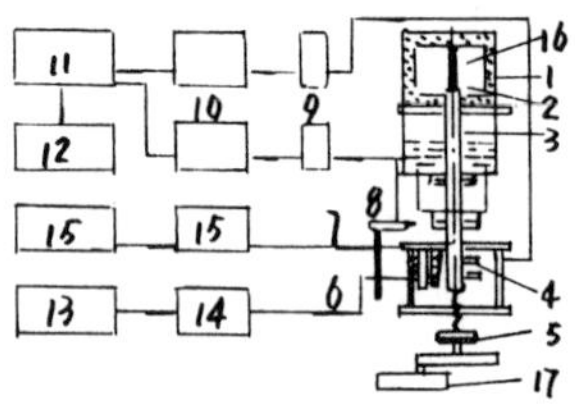

Figure 3 Test instruments

1 furnace, 2 specimen, 3 crystallizer, 4 accelerometer, 5 transfer machine, 6 magnetoelectric exciter, 7 hammer, 8 force transducer, 9 charge amplifiers, 10 voltmeter, 11 signal analyzer, 12 X–Y plotter, 13 signal generator, 14 power amplifier, 15 D. C. power supply, 16 crucible, 17 electromotor.

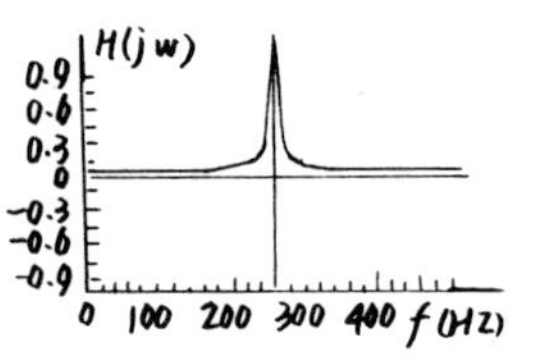

Figure 4 Frequency response function of the solidifying system

After exciting, sampling and analyzing, the position of the solid / liquid interface is measured directly by a piece of alundum probe inserting into the crucible and by a gauge. Thus, we get a pair of r and λ. Then all conditions are held constant except the level of cooling water. The solid phase of speciment gets longer when level raised to the desired height. Through meter monitoring, it is sure whether the steady thermal field is set up or not. As long as the field gets steady again, the experimental procedures will be repeated. Therefore, another pair of r and λ is obtained. By the repetition of the procedure. we acquire some pairs of r and λ. The experimental results and the computed results are listed in Table. 2. the comparison is shown in Fig.5.

The relationship between the position of the solid / liquid interface and the response power spectrum has been showed by Fig.6. In Fig.6, the response auto–spectrum is given which is correspondent to different position of the solid / liquid interface. The axis of abscissas represents frequency while the ordinate axis represents the solid length. The frequency coinciding with peak value of spectrum line is the eigenvalue correspondent to a given ratio of solid length to the total length. In practice, the map of spectral sequences at different moment can be recorded by X–Y plotter and used to locate the solid / liquid interface in real time.

TABLE 2 The comparison of the measured frequency ratio λ and natural frequency f with computed ones under different ratio of the length of solid phase to the total length

Test No.	r (length ratio)	λ Measured values	λ Computed values	f(Hz) Measured velues	f(Hz) Computed values
1	0.31	1.60	1.614	281.2	282.4
2	0.40	1.63	1.645	285.1	288.3
3	0.50	1.67	1.649	292.2	288.7
4	0.60	1.62	1.618	283.2	283.1
5	0.70	1.55	1.555	271.4	272.3
6	0.81	1.48	1.459	259.7	257.1
7	0.92	1.40	1.344	246.0	235.0

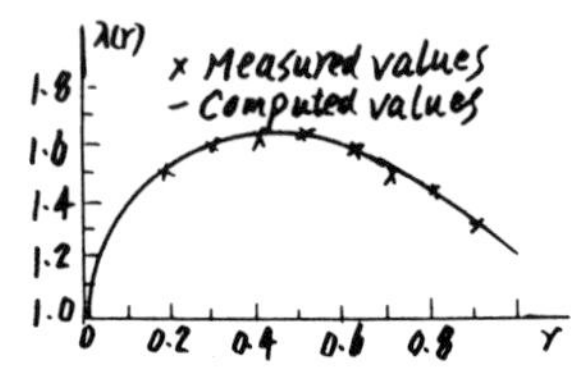

Figure 5 Measured values and computed values of $\lambda(r)$

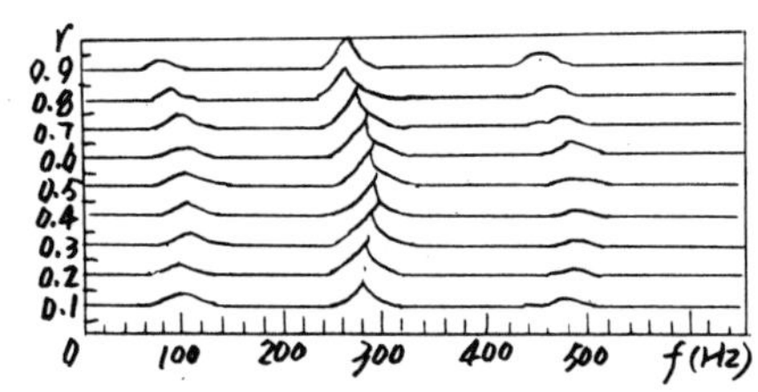

Figure 6 Relationship between the position of the solid / liquid interface and the response power spectrum

CONCLUSION

(1). It is demonstrated by experiment and calculation that the position of the solid / liquid interface can be determined by vibration parameters during solidification.

(2). In view of the characteristics of Bridgman solidifying system, authors adopt the sensitivity of transverse eigenvalue with respect to the position of the solid / liquid interface and locate the interface outside the furnace in real time.

(3). The work reported here gives the computed formula and relevant experimental detecting methods of sensitivity in Bridgman system. In identification, sensitivity may be "sensitive" or "blind".

REFERENCES

1. R. L. Paker, J.R. Manning and N.C. Peterson, Application of pulse—echo ultrasonics to locate the solid / liquid interface during solidfication and melting of steel and other metls, J. Appl. Phys, **58**(1985).
2. M.C. Flemings, Solidification Processing, McGram—Hill Book Company, (1974).
3. M.A. Haroun, Dynamics Analysis of Liquid Storage Tanks, EERL Rep, **4**(1980).
4. R.L. Fox, Optimization Methods for Engineering Design, Addison—Wesley Publishing Company, (1971).

DE-Vol. 63, Vibrations of Mechanical Systems
and the History of Mechanical Design
ASME 1993

THE USE OF CONDITION NUMBER OF A FRF MATRIX IN MECHANICAL PARAMETER IDENTIFICATION

Jhy-Horng Wang and Chen-Sung Liou
Department of Power Mechanical Engineering
National Tsing Hua University
Hsinchu, Taiwan

ABSTRACT

A new identification algorithm is proposed in this work to identify the parameters of mechanical joints. The method considers the whole structure as two substructures which are connected by the joints to be identified. The frequency response functions of the whole structure and substructures are used to extract the joint parameters. In contrast to the traditional methods, only a FRF matrix is needed to inverse in the proposed method. Therefore, it is possible to calculate the condition number of the FRF matrix before the identification. The condition number is defined as the ratio of the maximum singular value divided by the minimum one of a matrix. The condition number of noise contaminated FRF matrix can be used to indicate the sensitivity of the FRF matrix to measurement noise. Therefore, the condition number can be used to avoid the ill–conditioned problem by eliminating the ill–conditioned FRF in some frequency ranges before identification. The simulated results show that the proposed method can significantly improve the accuracy of identification.

INTRODUCTION

The accuracy of the mathematical model of a mechanical system depends on the accuracy of the parameters of the model. Most of the mechanical parameters are very difficult to determine by theoretical methods. A number of methods have been proposed in the past to identify the mechanical parameters experimentally. The most trobulesome problem encountered in experimental parameter identification is the effect of measurement noise. The problem caused by measurement noise has been widely discussed by many works. Fritzen (1986) and Wang (1988) have applied the instrumental variable method to minimize the estimation bias caused by measurement noise. However, it lacks a general rule to find a proper instrumental variable series. A sequential frequency domain filter has been proposed by Mottershead and his colleagues (1986, 1988, 1990) to overcome the bias problem in the presence of measurement noise. An adaptive kalman filter has been proposed by Hac (1990) to improve the quality of the noisy data. However, the method requires a good initial guess of the system parameters, and an iterative procedure is necessary. The singular value decomposition method and the Eigensystem Realization Algorithm have been widely applied to minimize the measurement noise effect for "nonstructural" parameter identification (Kung, 1978; Juang and Pappa, 1986; Elliott, et. al., 1988).

Among the structural parameters, the parameters of mechanical joints (e.g. sliding joint, bolted joint) are most difficult to find by theoretical methods. Yuan (1985) and Kim (1989) used the finite element method combined with the DDS method to identify the joint properties of machine tools. However, some modeling errors may be induced in the finite element model. A method based on the substructure synthesis has been proposed by Tsai (1988) to identify the properties of mechanical joints. The method has some advantages, for instance, it does not require the mathematical model of the structue. However, the method is very sensitive to the unavoidable measurement noise, a small noise level may cause the identified result to be faulty. A method was proposed by Wang (1990, 1991) to overcome the problem of measurement noise. However, the method requires that the frequency response functions at the joint interfaces of the whole structure should be measurable. In some mechanical joints, the joint interfaces are inaccessible, and thus the frequency response functions (FRF) at the joint interfaces are not measurable.

In this work, a method was proposed to identify the parameters of mechanical joints which are inaccessible by vibration sensors. Because the measurement noise is unavoidable in practice, the noise problem was especially considered in this work. Only a single FRF matrix was needed to inverse in the proposed method. Therefore, it is possible to calculate the condition number of the special FRF matrix before identification. The condition number can then be used to indicate the sensitivity of the FRF matrix to the measurement noise.

THEORETICAL FORMULATION

A whole structure generally can be devided into two substructures from the interfaces of the joint to be identified. It is assumed that the frequency of interest is not too high, the mass of the joint can be neglected and the joint can be modeled as linear spring and damper elements, as shown in Fig. 1. The objective of the proposed method is to identify the joint properties experimentally from the frequency response functions (FRFs) of the whole structure and the substructures.

With the definition of FRFs, the relations between the displacement vectors and the force vectors of substructures 1 and 2, (see Fig. 1), can be written as

$$\begin{Bmatrix} \{X_e\}_1 \\ \{X_a\}_1 \end{Bmatrix} = \begin{bmatrix} [H_{ee}]_1, [H_{ea}]_1 \\ [H_{ae}]_1, [H_{aa}]_1 \end{bmatrix} \begin{Bmatrix} \{F_e\}_1 \\ \{F_a\}_1 + \{F_j\}_1 \end{Bmatrix} \tag{1}$$

$$\begin{Bmatrix} \{X_b\}_2 \\ \{X_c\}_2 \end{Bmatrix} = \begin{bmatrix} [H_{bb}]_2, [H_{bc}]_2 \\ [H_{cb}]_2, [H_{cc}]_2 \end{bmatrix} \begin{Bmatrix} \{F_b\}_2 + \{F_j\}_2 \\ \{F_c\}_2 \end{Bmatrix} \tag{2}$$

where $\{X_a\}_1$ and $\{X_b\}_2$ represent the displacement vectors on the joint interfaces of substructures 1 and 2, as shown in Fig. 1; $\{X_e\}_1$ and $\{X_c\}_2$ represent the displacement vectors on all other regions except the joint interfaces of substructures 1 and 2. The vectors $\{F_a\}_1$ and $\{F_e\}_1$ represent the external force vectors acting on the joint interface and other region of the substructure 1, while the vectors $\{F_b\}_2$ and $\{F_c\}_2$ represent the external force vectors acting on the joint interface and other region of the substructure 2. The joint force vectors are represented by $\{F_j\}_1$ and $\{F_j\}_2$, and they are equal in magnitude, but opposite in direction, i.e.,

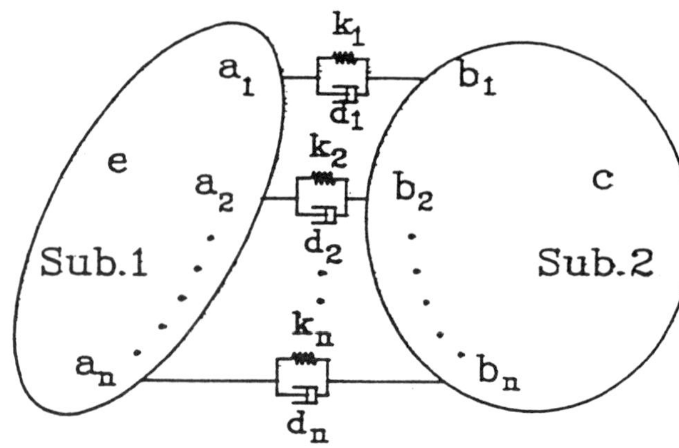

FIG.1 A STRUCTURE INCLUDING TWO SUBSTRUCTURES CONNECTED BY THE TO BE IDENTIFIED JOINTS k_1, d_1, k_2, d_2,...k_n, d_n.

$$\{F_j\}_1 = -\{F_j\}_2 \tag{3}$$

The displacement vectors at the joint interfaces are related to the joint force by

$$\{X_b\}_2 - \{X_a\}_1 = [H_j] \cdot \{F_j\}_1 \tag{4}$$

with

$$[H_j] \equiv [P_j]^{-1} = \begin{bmatrix} k_1 + j\omega d_1, & 0, & \cdots & 0 \\ 0, & k_2 + j\omega d_2, & \cdots & 0 \\ \vdots & & \ddots & \vdots \\ \cdots & & \cdots & k_n + j\omega d_n \end{bmatrix}^{-1}$$

where $j = \sqrt{-1}$ and $k_1, k_2 \ldots k_n, d_1, d_2 \ldots d_n$ are the constant spring and damping coefficients of the joint, as shown in Fig. 1. Substitute equations (3) and (4) into equations (1) and (2), the relations between the displacement vector and external force vector of the whole structure can be obtained (Wang, 1990),

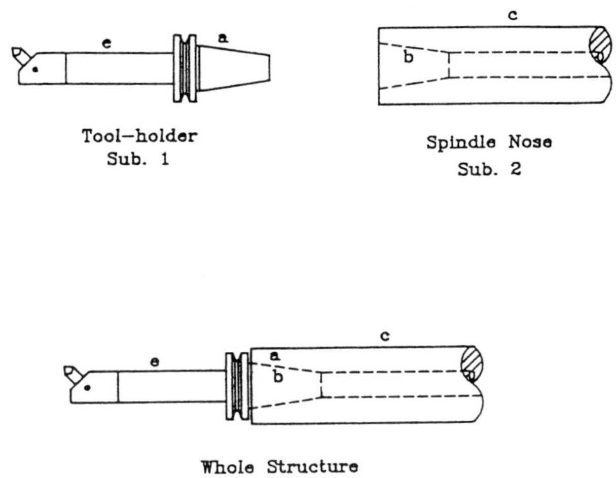

FIG.2 TOOL–HOLDER AND SPINDLE NOSE. THE TAPER INTERFACES OF THE WHOLE STRUCTURE ARE INACCESSIBLE BY VIBRATION SENSOR.

$$\begin{Bmatrix} \{X_e\}_1 \\ \{X_a\}_1 \\ \{X_b\}_2 \\ \{X_c\}_2 \end{Bmatrix} = \begin{bmatrix} [H_{ee}], [H_{ea}], [H_{eb}], [H_{ec}] \\ [H_{ae}], [H_{aa}], [H_{ab}], [H_{ac}] \\ [H_{be}], [H_{ba}], [H_{bb}], [H_{bc}] \\ [H_{ce}], [H_{ca}], [H_{cb}], [H_{cc}] \end{bmatrix} \begin{Bmatrix} \{F_e\}_1 \\ \{F_a\}_1 \\ \{F_b\}_2 \\ \{F_c\}_2 \end{Bmatrix} \tag{5}$$

The FRF matrix of the whole structure in equation (5) are expressed in terms of the FRF matrices of the substructures and the joint properties.

In the previous work (Wang, 1990), it was assumed that all the FRF matrices in equation (5), i.e., $[H_{ee}]$, $[H_{ea}]$, $[H_{eb}]$ ... $[H_{cc}]$, were measurable. Generally, this assumption is reasonable. However, in some cases, the joint interfaces (i.e. regions "a", and "b" in Fig. 1) of the whole structure are inaccessible. As a consequence, the corresponding FRF matrices in this regions are not measurable. For instance, one wants to identify the joint parameters of a taper joint used in the tool–holder of a machine tool spindle, as shown in Fig. 2. If the tool–holder and the spindle are considered as a whole structure, then it is impossible to measure the vibration at the taper joint interfaces of the whole structure (regions "a", "b" in Fig. 2). That means, the matrices, $[H_{ea}]$, $[H_{eb}]$, $[H_{ea}]$, $[H_{ab}]$, $[H_{ac}]$ .... $[H_{ca}]$, $[H_{cb}]$, are not measurable, only the four matrices, $[H_{ee}]$, $[H_{ec}]$, $[H_{ce}]$ and $[H_{cc}]$ are available. Therefore, two points are especially emphasized in this work:

(1) The joint interfaces of the whole structure are inaccessible.

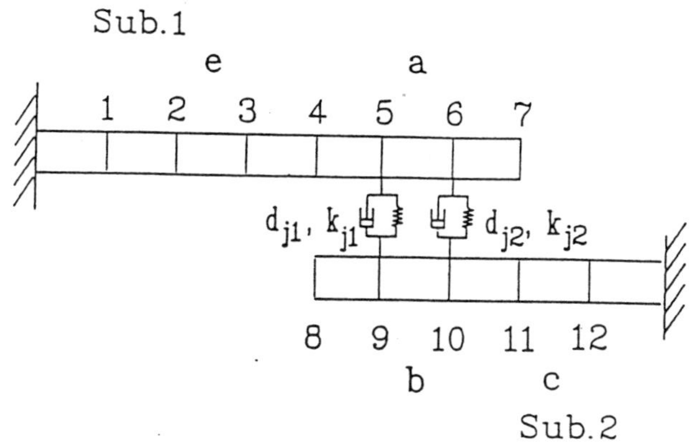

FIG.3 MODEL OF THE SIMULATED EXAMPLE.

(2) A new identification method is introduced to overcome the problem caused by the measurement noise in the FRFs.

The available matrices, $[H_{ee}]$, $[H_{ec}]$, $[H_{ce}]$ and $[H_{cc}]$ can be expressed in terms of the FRF matrices of the substructures and the joint properties,

$$[H_{ee}] = [H_{ee}]_1 - [H_{ea}]_1 [H_B]^{-1}[H_{ae}]_1 \tag{6}$$

$$[H_{ec}] = [H_{ea}]_1[H_B]^{-1}[H_{bc}]_2 \tag{7}$$

$$[H_{ce}] = [H_{cb}]_2[H_B]^{-1}[H_{ae}]_1 \tag{8}$$

$$[H_{cc}] = [H_{cc}]_2 - [H_{cb}]_2 [H_B]^{-1}[H_{bc}]_2 \tag{9}$$

with $[H_B] = [H_{aa}]_1 + [H_{bb}]_2 + [H_j]$

One can find that all the four equations (6) – (9) contain the joint parameters $[H_j]$, so theoretically each equation can be used to extract the joint parameters. In fact, equations (6) – (9) can be written in a general form as:

$$[H_\gamma] = [H_\alpha][H_B]^{-1}[H_\beta] \tag{10}$$

In the work by Tsai (1988), an identification algorithm was developed from equation (10) to identify the parameter of bolt joint. For comparison with the new method proposed in this work, the algorithm by Tsai (1988) was given here again for reference. Taking inverse at the both sides of equation (10), one has:

$$[H_\gamma]^{-1} = [H_\beta]^{-1}([H_{aa}]_1 + [H_{bb}]_2 + [H_j])[H_\alpha]^{-1}$$

or

$$[H_\gamma]^{-1} - [H_\beta]^{-1}([H_{aa}]_1 + [H_{bb}]_2)[H_\alpha]^{-1} = [H_\beta]^{-1}[H_j][H_\alpha]^{-1} \quad (11)$$

Taking inverse at the both sides of equation (11), one has,

$$\left[[H_\gamma]^{-1} - [H_\beta]^{-1}([H_{aa}]_1 + [H_{bb}]_2)[H_\alpha]^{-1}\right]^{-1} = [H_\alpha][H_j]^{-1}[H_\beta] = [H_\alpha][P_j][H_\beta] \quad (12)$$

If the FRFs are known (for example by measurement), then the only unknowns in equation (12) are the joint parameters $[P_j]$, and the least squares method (Wang, 1990) can be used to obtain the joint parameters.

It can be found that the derivation of equation (12) is straight forward from the mathematical point of veiw. Theoretically, if the values of the FRFs are exact, one can obtain the exact joint parameters form equation (12) because there is no mathematical approximation in deriving equation (12). However, one may say that the derivation of equation (12) is correct only from a pure mathematical point of view because it does not consider any possible error (for instance, measurement noise) in FRFs. In practice, a small noise or measurement error in the FRFs will cause the identified joint parameters to deviate from the exact values drastically. The reason for this trouble is due to the fact that there are many inverse operations on the FRF matrices in equation (12), and the FRF matrices may become ill-conditioned in some frequency ranges.

To overcome this problem, a new identification algorithm is proposed here. The idea of the proposed method basically is based on the understanding of the problem caused by the unavoidable measurement noise.

From equation (10), let

$$[H_B]^{-1}[H_\beta] \equiv [Y] \quad (13)$$

then

$$[H_\gamma] = [H_\alpha][H_B]^{-1}[H_\beta] = [H_\alpha][Y] \quad (14)$$

Since $[H_\gamma]$ and $[H_\alpha]$ can be known by measurement, one can solve $[Y]$ matrix by

$$[Y] = [H_\alpha]^{-1}[H_\gamma] \quad (15)$$

From the definition of $[Y]$ in equation (13), one has,

$$[Y] = [H_B]^{-1}[H_\beta] = ([H_{aa}]_1 + [H_{bb}]_2 + [H_j])^{-1}[H_\beta]$$

or

$$([H_{aa}]_1 + [H_{bb}]_2 + [H_j])[Y] = [H_\beta]$$

$$[H_j][Y] = [H_\beta] - ([H_{aa}]_1 + [H_{bb}]_2)\,[Y] \quad (16)$$

With the definition of $[H_j]$ in equation (4),equation (16) becomes

$$[Y] = [P_j]\{[H_\beta] - ([H_{aa}]_1 + [H_{bb}]_2)\,[Y]\}$$

or

$$\{[H_\beta] - ([H_{aa}]_1 + [H_{bb}]_2\,[Y]\}^T[P_j] = [Y]^T \quad (17)$$

Since the $[Y]$ matrix is already known from equation (15) and other FRF matrices can also be known by measurement, the only unknowns in equation (17) are the joint parameters $[P_j]$. The joint parameters can then be identified by the least squares method.

Not like th derivation of equation (12), the derivation of equation (17) is somewhat zigzag from the mathematical point of view. The goal of this mathematical formulation is to minimize the noise effect by reducing the matrix inverse operations. Note that the only matrix inverse operation in equation (17) is $[H_\alpha]^{-1}$ (see equation (15)). If the $[H_\alpha\}$ matrix is ill-conditioned, then the parameters identified by equation (17) may be faulty. Because the FRF matrix is a function of frequency ω, generally the FRF matrix becomes ill-conditioned only in some frequency ranges. So, the concept of condition number of a matrix was adopted in this work to find the frequency ranges in which the $[H_\alpha]$ matrix becomes ill-conditioned.

The ILL-Conditioned Problem

There is only an inverse operation on the FRF matrix in the proposed identification equation (i.e., equation (17)), so, it is possible to avoid the ill-conditioned problem by eliminating some unwanted data in the FRF matrix. Note that all the FRF matrices are function of frequency ω. In some frequency ranges, the $[H_\alpha]$ matrix may become ill-conditioned. If the FRF data in these frequency ranges are not eliminated, then the identified results may be faulty. So, the first step is to find the frequency ranges in which the $[H_\alpha]$ becomes ill–conditioned. The concept of condition number of a matrix is adopted here to find the ill–conditioned frequency ranges.

Consider a linear system

$$[A]\{x\} = \{b\} \tag{18}$$

where $\{b\} \in R^n$ and $\{A\} \in R^{n\times n}$ is nonsingular. We suppose that $[A]$ and $\{b\}$ are changed to $[A] + \Delta[A]$ and $\{b\} + \Delta\{b\}$ by a small perturbation $\Delta[A]$, $\Delta\{b\}$. The perturbed system satisfies:

$$([A] + \Delta[A])\{y\} = \{b\} + \Delta\{b\} \tag{19}$$

It can be proved that (Lancaster, 1985):

$$\frac{\|\{x\}-\{y\}\|}{\|\{x\}\|} \leq \frac{\kappa(A)}{1-(\kappa(A)\|\Delta A\|/\|A\|)}\left(\frac{\|\Delta\{b\}\|}{\|\{b\}\|} + \frac{\|\Delta A\|}{\|A\|}\right) \tag{20}$$

where $\| \ \|$ denotes matrix or vector 2–norm, and $\kappa(A) = \|A\| \|A^{-1}\|$, defined as the condition number of A. Equation (20) shows the upper bound for the relative perturbation of $\{x\}$ in terms of the relative perturbations of $\{b\}$ and $[A]$ and the condition number $\kappa(A)$. If the relative perturbation of $\{b\}$ and $[A]$ are small, then the relative perturbation of $\{x\}$ is determined by the condition number of $[A]$.

Let us return to our original problem, i.e., equation (15). The $[H_\alpha]$ and $[H_\gamma]$ are measured frequency reponse functions. Generally, they are contaminated by measurement noise. Because $[H_\alpha]$ is function of frequency ω, the condition numbers of $[H_\alpha]$ are also function of frequency ω. According to equation (20), if the condition number of $[H_\alpha]$ becomes very large in some frequency ranges, then the $[Y]$ matrix in equation (15) may have large error due to a small noise in $[H_\alpha]$. Note that the $[H_\alpha]$ is a complex matrix, its condition number is more complicated than that of a real matrix. For a complex FRF matrix $[H_\alpha]$, one may write that:

$$[H_\alpha] = [P] + j[Q] \tag{21}$$

where $[P]$ and $[Q]$ are the real and imaginary parts of $[H_\alpha]$ respectively. If let $[H_\alpha]^{-1}$ equal to $[R] + j[S]$, then one has:

$$([P] + j[Q])([R] + j[S]) = [I] \tag{22}$$

Solve equation (22), $[R]$ and $[S]$ can be written in terms of $[P]$ and $[Q]$ as:

$$[R] = ([P] + [Q][P]^{-1}[Q])^{-1} \tag{23}$$

$$[S] = -([Q] + [P][Q]^{-1}[P])^{-1} \tag{24}$$

From equation (23) and (24), one can know that the $[H_\alpha]$ matrix is well–conditioned only if the matrices $[P]$, $[Q]$, $[P]+[Q][P]^{-1}[Q]$ and $[Q] + [P][Q]^{-1}[P]$are well–conditioned. If the condition numbers of the four matrices, say $\kappa([P])$, $\kappa([Q])$, $\kappa([P]+[Q][P]^{-1}[Q])$and $\kappa([Q] + [P][Q]^{-1}[P])$ are "small enough", then one may say that the complex matrix $[H_\alpha]$ is well–conditioned. However, there are no definite rules to decide "how small" the condition numbers should be in order to ensure that the matrix is well–conditioned. Because the FRF matrix is function of frequency, so is the condition number. In this work, the condition numbers of the above mentioned four matrices are first calculated in the frequency range of interest. The average values of the condition numbers can then be found. If the condition numbers of the four matrices are larger than their average values in some discrete frequencies, then the data of $[H_\alpha]$ in these discrete frequencies are eliminated and are not used for identification .

RESULTS OF NUMERICAL SIMULATION

Note that if there is no noise in the FRF matrices, the joint parameters can be identified exactly by either equation (12) or (17), because there is no mathematical approximation in the derivations of these equations. However, measurement noise is always unavoidable in practice. Therefore, in the following simulation, noise was added to the FRF matrices to simulate the situation in practice.

Form of Noise.

A random noise with Gaussian distribution (zero mean, variance σ_n^2) was added to the FRFs to simulate the measurement noise. If $H_{ij}(\omega)$ represents the frequency response function between the ith and jth points of a structure, then the noise level E is defined as:

$$E^2 = \frac{\sigma_n^2}{|H_{ij}(\omega)|^2_{max}} \tag{25}$$

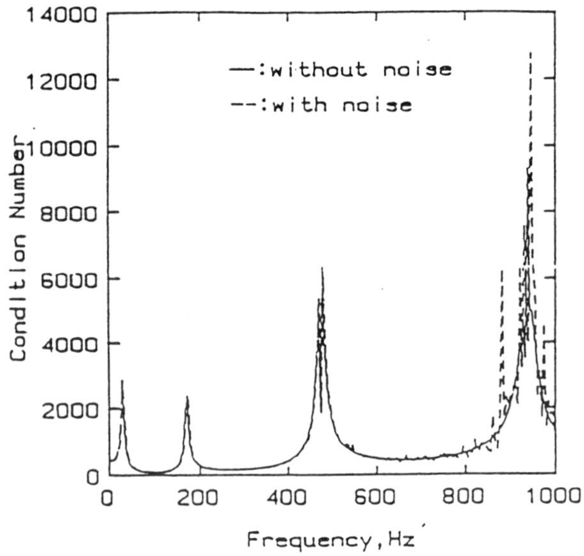

FIG.4 THE CONDITION NUMBER OF THE REAL PART OF $[H_\alpha]=[H_{ea}]_1$. THE CONDITION NUMBER OF THE NOISE CONTAMINATED $[H_\alpha]$ IS ALSO SHOWN.

where $|H_{ij}(\omega)|_{max}$ is the maximum absolute value of $H_{ij}(\omega)$ in the frequency range of interest. If the noise level is given, then a set of random noise can be generated by a computer program. Note that the frequency response function is a complex value, the random noise should be added to the real and imaginary parts of the FRFs respectively.

Simulated Example.

The simulated system consists of two parallel beams connected together through two bolted joints. The model of the system is shown in Fig. 3. The whole system is approximated by twelve finite element with twelve degrees of freedom, i.e., only one degree of freedom is considered in each node point. The bolted joints are modeled as two linear spring and damping elements, indicated by k_{j1}, k_{j2}, d_{j1} and d_{j2} in Fig. 3. If the joint parameters k_{j1}, k_{j2}, d_{j1}, and d_{j2} should be identified, then the whole strucute is divided into two substructures from the joints. The maximum frequency, f_{max}, was set to be 1000 Hz. The theoretical FRFs are continuous function of frequency, however, only discrete FRFs with finite resolution can be obtained by measurement. Therefore, a frequency resolution of the FRFs, Δf, was set to be 5Hz to simulate the practical situation. The detailed procedure of the simulation is summarized here:

(1) Generate the necessary FRF matrices from the equation of whole structure and the substructures.

(2) Determine the number $N = f_{max}/\Delta f$, then N discrete values with Gaussian distribution were generated and added to the discrete FRFs to simulate the noise.

(3) The condition numbers of $[H_\alpha]$ at each discrete frequency were calculated, and an average value of the condition numbers was found.

(4) Only the $[H_\alpha]$ with condition numbers smaller than the average value were kept to identify the joint parameters.

In this example, the number N was set to be 200 = 1000 Hz/5Hz, and the noise level was set to be 2%. For comparison, equation (12) and equation (17) without considering the condition number were also used to identify the joint parameters.

Note that the $[P_j]$ matrix in equation (17) is with order nxn, where n is the degrees of freedom of joints. If all the other matrices in equation (17) are also with order nxn, then there are adequate equations to solve the unknowns in $[P_j]$ matrix. In other words, the number of the measurement points on regions "e" and "c" (see Fig. 1) should be taken the same as the degrees of the freedom of the joints in order to ensure that all the FRF matrices in equation (17) are square matrix with order nxn. In this simulated examples, the order of the matrix $[P_j]$ is 2×2. As shown in Fig. 3, the degrees of freedom on region "e" are four, marked by 1, 2, 3, 4; and the degrees of freedom on region "c" are two , marked by 11, and 12. In the following discussions, the points 2 and 3 in region "e" and points 12, and 13 in region "c" were chosen as the measurement points.

The joint parameters identified from equation (12) are listed in Table 1. One can find that the error of the identified results is very large, and the values of k_{j1} and d_{j2} even become negative. The reason for this large error is due to many inverses on the FRF matrices in equation (12). It is well known that the FRF matrix may become ill-conditioned in some discrete frequencies. That means a small noise in the FRF matrix may cause a large error in the inversed matrix.

The results identified by equation (17) without considering the condition numbers are shown in Table 2. One can find that the identified results are more accurate than the results in Table 1. This is due to the fact that there is only an inverse operation on FRF matrix in equation (17). The error was mainly due to the ill-conditioned $[H_\alpha]$ in some discrete frequencies. The condition number of the real part of $[H_\alpha]$ is shown Fig 4. One can find that the condition numbers are especially large in some narrow frequency ranges. According to the proposed method, the data of $[H_\alpha]$ at these narrow

TABLE 1 IDENTIFIED RESULT BY USING EQUATION (12)

Properties		Exact values	Identified values	Errors
Stiffness (N/m)	k_{J1}	120000	-390240.5	-425.2%
	k_{J2}	90000	406299.8	351.4%
Damping (N-s/m)	d_{J1}	34	511.2	1403%
	d_{J2}	32	-116.5	-464.2%

TABLE 2 IDENTIFIED RESULT BY USING EQUATION (17) WITHOUT CONSIDERING THE CONDITION NUMBER.

Properties		Exact values	Identified values	Errors
Stiffness (N/m)	k_{J1}	120000	97871.6	-18.4%
	k_{J2}	90000	78912.8	-12.3%
Damping (N-s/m)	d_{J1}	34	35.3	4.0%
	d_{J2}	32	25.3	-20.9%

frequency ranges should be eliminated to ensure that the $[H_\alpha]^{-1}$ is well-conditioned. As mentioned, there are no definite rules to decide "how small" the condition numbers should be in order to ensure that the matrix is well-conditioned. In this work, the average value of the condition numbers in the frequency range of interest was chosen as a limit. If the condition number is smaller than the average value, the matrix is regarded as well-conditioned. The results identified by equation (17) with the consideration of the condition number is listed in Table 3. One can find that the error of the result is only about the same order as the noise level. The result indicated that the troublesome problem caused by measurement noise can be overcome by the proposed method.

TABLE 3 IDENTIFIED RESULT BY THE PROPOSED METHOD, EQUATION (17) WITH THE CONSIDERATION OF THE CONDITION NUMBER.

Properties		Exact values	Identified values	Errors
Stiffness (N/m)	k_{J1}	120000	116345.8	-3.0 %
	k_{J2}	90000	90083.1	0.1 %
Damping (N-s/m)	d_{J1}	34	32.3	-4.9 %
	d_{J2}	32	31.2	-2.5 %

CONCLUSIONS

A parameter identification method was introduced in this work. The method used the measured FRFs to identify the parameters of inaccessible joints. As pointed out by some works, the unavoidable measurement noise and the ill-conditioned problem of the FRFs are the major difficulty when the inversed FRFs are used to identify the mechanical parameters. In this work, the noise problem was overcome by a new identification algorithm in which only a FRF matrix should be inversed. So, it becomes possible to use the condition number of the FRF matrix to ensure that the FRF matrix is well-conditioned in the selected frequencies. The results show that the error of the identification is only about the same order of magnitude as the measurement noise level.

REFERENCES

Elliott, K.B., Juang, J.N., and Robinson, J., 1988, "Force Prediction Using Singular-Value-Decomposition," 6th International Modal Analysis Conference, Florida, pp. 1582 – 1588.

Fritzen, C.-P., 1986, "Identification of Mass, Damping, and Stiffness Matrices of Mechanical Systems," ASME Journal of Vibration, Acoustics, Stress and Reliability in Design, Vol. 108, pp. 9–16.

Hac, A. and Spanos, P.D., 1990, "Time Domain Method for Parameter System Identification," ASME Journal of Vibration and Acoustics, Vol. 112, pp. 281–287.

Juang, J.N., and Pappa, R.S., 1986, "Effect of Noise on Modal Parameter Identification by the Eigensystem Realization Algorithm," Journal of Guidance, Control, and Dynamics, Vol. 9. pp. 294–303.

Kim, T.R., Wu, S.M., and Eman, K.F., 1989, "Identification of Joint Parameters of a Taper Joint," ASME Journal of Engineering for Industry, Vol. 111, pp. 282–287.

Kung, S., 1978, "A New Identification and Modal Reduction Algorithm Via Singular Value Decomposition," 12th Asilomar Conference on Circuits, Systems and Computers, pp. 705–714.

Lancaster, P. and Tismenetsky, M., 1985, "The Theory of Matrices," Second Edition, Academic Press.

Mottershead, J.E., and Stanway, R., 1986, "Identification Structural Vibration Parameters by Using a Frequency Domain Filter," Journal of Sound and Vibration, Vol. 109, pp. 495–506.

Mottershead, J.E., 1988, "A Unified Theory for Recursive Frequency Domain Filters and Application to System Identification in Structural Dynamics," ASME Journal of Vibration, Acoustics, Stress, and Reliability in Design, Vol. 110, pp. 360–365.

Mottershead, J.E., Tee, T.K., Foster, C.D., and Stanway, R., 1990, "An Experiment to Identify the Structural Dynamics of a Portal Frame," ASME Journal of Vibration and Acoustics, Vol. 112, pp. 78–83.

Tsai, J.S., and Chou, Y.F., 1988, "The Identification of Dynamic Characteristics of a Single Bolt Joint," Journal of Sound and Vibration, Vol. 125, pp. 487–502.

Wang, J.H., 1988, "Mechanical Parameters Identifictaion with Special Consideration of Noise Effects," Journal of Sound and Vibration, Vol. 125, pp. 151–167.

Wang, J.H., and Liou, C.M., 1990, "Identification of Parameters of Structural Joints by Use of Noise–Contaminated FRFs," Journal of Sound and Vibration, Vol. 142., pp. 261–277.

Wang, J.H. and Liou, C.M., 1991, "Experimental Identification of Mechanical Joint Parameters," ASME Journal of Vibration and Acoustics, Vol. 113, pp. 28–36.

Yuan, J.X., and Wu, X.M., 1985, "Identification of the Joint Structural Parameters of Machine Tool by DDS and FEM," ASME Journal of Engineering for Industry, Vol. 107, pp. 64–69.

DE-Vol. 63, Vibrations of Mechanical Systems
and the History of Mechanical Design
ASME 1993

MEASUREMENT OF STATIC AND DYNAMIC PROPERTIES OF A TILTING PAD BEARING AND COMPARISON WITH THEORY

Thomas Bonner and John H. Vohr
Steam Turbine Engineering
General Electric Company
Schenectady, New York

ABSTRACT

Measurements were made of the static and dynamic properties of a seven inch diameter, four-pad, tilting pad bearing with both centrally pivoted and 60% offset pads. Data was recorded at 1500, 1800 and 3000 RPM and covered a range of Sommerfeld numbers from .055 to .625. Dynamic properties were measured by applying unbalance weights of various magnitudes to the rotor and measuring the synchronous response amplitudes and phase angles. Good agreement was obtained between measured properties and those predicted using linearized stiffness and damping coefficients obtained from isoviscous lubrication theory.

INTRODUCTION

The static and dynamic testing of tilting pad bearings was motivated by the effort to optimize the rotor dynamic characteristics of a particular generator design. For this generator, calculations predicted that the damping at critical speeds would be significantly enhanced if offset pivot bearings were used instead of the standard, centrally pivoted design. Since the use of offset pivot bearings was an uncommon practice, it was decided to conduct a bearing test to evaluate the difference in the dynamic characteristics of otherwise identical 0.5 and 0.6 pivot ratio bearings. The results of this testing and the comparison of results with theoretical predictions form the content of this paper.

DESCRIPTION OF TEST BEARING

The bearings tested were designed and manufactured by Kingsbury Incorporated. They consisted of a complete bearing assembly with 0.5 pivot pads plus four replacement pads with a 0.6 pivot ratio. The test bearing configuration with the 0.5 pivot ratio pads is shown in Figure 1. The geometrical parameters for the test bearing are given in Table 1.

TABLE 1 - GEOMETRICAL PARAMETERS FOR TEST BEARINGS

Diameter	7"
Length	5"
Radial Clearance	.0055"
Number of Pads	4
Configuration	Load between pad
Pad Circumferential Length	70°
Pivot Location	0.5 and 0.6
Preload	0
Test Oil Viscosity	19.6 cs at 100° F

DESCRIPTION OF TESTING SYSTEM

Test Machine

The test machine used was built to test steady and dynamic performance of 7" diameter journal bearings. A schematic diagram of this machine is shown in Figure 2.

A single test bearing is used on the machine. The shaft of the machine is approximately four feet long and very stiff. Stub shafts are bolted to each end of the main shaft. The stub shafts contain a series of holes which will accept unbalancing weights to apply a synchronous dynamic force.

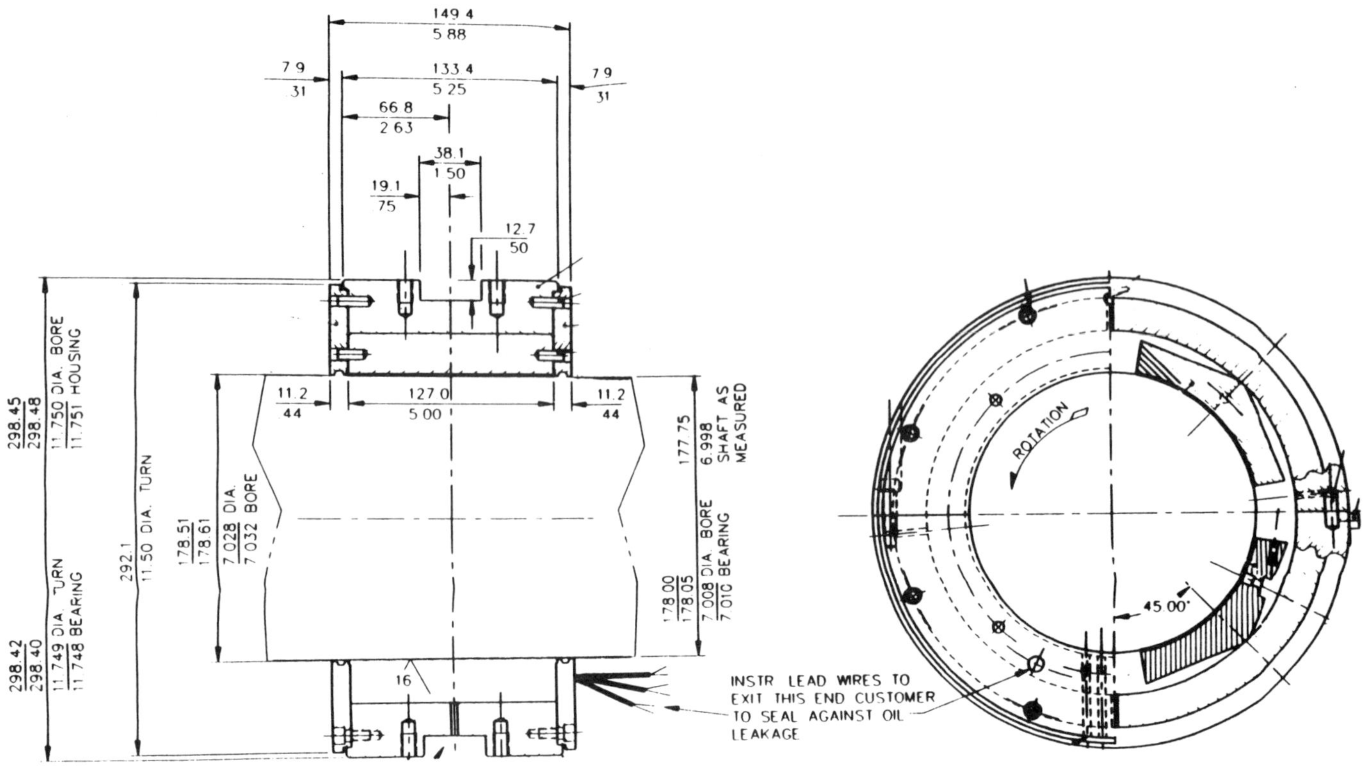

FIGURE 1 TEST BEARING WITH 0.5 PIVOT RATIO PADS

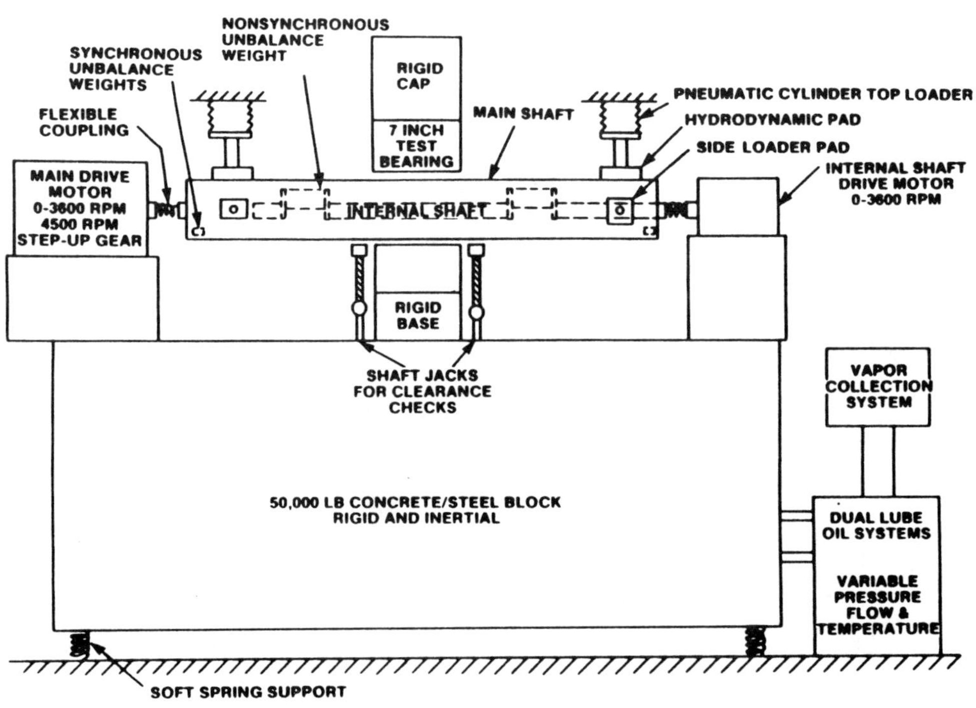

FIGURE 2 SCHEMATIC OF TEST MACHINE

The shaft is driven by a variable speed, 40 HP, 3600 RPM DC motor. The shaft is coupled to the motor by a double universal joint coupling. This coupling is used to allow the rotor complete vibrational freedom in responding to the dynamic unbalance forces.

Two vertical steady state loaders are used in this test. The vertical force for each loader can be varied from zero to 4,000 pounds. The load is developed in flexible rubber pneumatic cylinders and applied to the shaft with an oil lubricated pad. The load is measured and digitally output by load cells which are placed in line with the loader pads. The low stiffness of the pneumatic cylinder (500 lbs/in) relative to the test bearing oil film (10^6 lbs/in) insures that the shaft dynamic motion will be negligibly affected by the steady load applicators. The small mass of the load pads (20-30 lbs) relative to the shaft mass allows their dynamic effects to be neglected.

Instrumentation

There are eight proximity probes used to determine the instantaneous shaft position. These probes are mounted on both test bearing seal rings at the locations shown in Figure 3. Two probes are sufficient to locate the shaft in the X-Y reference system, but having four allows the signals to be averaged. This eliminates the effects of thermal growth of the bearing and shaft on the displacement measurements. It is necessary to locate probes on either side of the bearing in order to determine the position of the axial mid-plane of the test rotor.

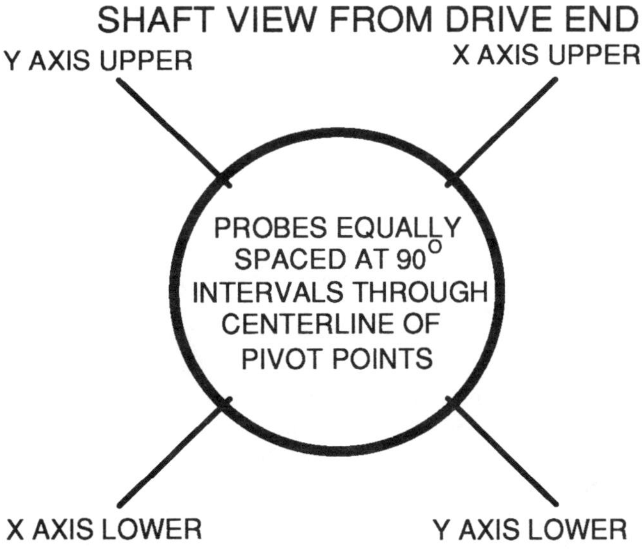

FIGURE 3 PROXIMITY PROBE LOCATIONS

In addition to the eight probes around the shaft, there is a probe mounted near the coupling. This probe is used to read the phase reference mark which is machined into a coupling plate. This phase mark is used as the reference for reading phase angles during dynamic testing.

The probes are individually connected to isolation amplifiers. The eight shaft probes are connected to differential amplifiers which are used to add the signals of the upper and lower probes. The signals from the differential amplifiers and the phase probe are connected directly to a digital vector filter (DVF). The digital vector filter is a two channel instrument that measures the amplitude and phase of a filtered vibration signal. When operating in the filtered mode, the vibration input is fed through a narrow band-width band-pass filter. The filter is tuned to the synchronizing signal so that only the component of vibration associated with the frequency of the synchronizing signal is extracted and displayed.

By using separate leads, the probes from the isolation amplifiers can be input directly into the DVF. This allows reading the displacement and phase angle at each probe.

THEORETICAL ANALYSIS

The theoretical predictions of linearized stiffness and damping coefficients for the test bearings were based on solutions to the two-dimensional Reynolds equation assuming constant viscosity in the lubricant film. The solutions were obtained numerically by a finite difference technique. The procedure for calculating stiffness and damping coefficients followed that described by Lund in Reference 1. Pad inertia was neglected. The stiffness and damping coefficients for each pad were modified to take account of the pivot stiffness. The viscosity used to evaluate the bearing film pressures, both static and dynamic, was the viscosity corresponding to the temperature measured three quarters of the circumferential distance along the centerline of each pad. For the bottom two pads, these measured temperatures agreed quite closely with each other and also with theoretically predicted values.

DISCUSSION OF TEST RESULTS AND COMPARISON WITH THEORY

Steady State Tests

Figures 4 through 7 show the vertical height of the rotor above bottom dead center (BDC) as measured by the sum of the readings obtained from the four y-axis proximeter probes and also from the four x axis probes (see Figure 3). As can be seen, the measured vertical height of the rotor, as determined from the four x displacement probes, agreed with that determined by y probes indicating that the shaft position was essentially in the vertical plane above BDC as theory would predict.

Measurements of vertical film height were obtained by increasing applied load from 1000 lbs to 7000 lbs in 1000 lb increments and then reducing load back to 1000 lbs also in 1000 lb increments. Actual total load on the bearing consisted of the applied load plus 500 lbs contributed by the rotor weight. The data obtained from the measurements made during the ascending load runs agreed very closely with that obtained during descending load runs. Hence, for clarity in presentation, only the data obtained during the descending load runs are plotted.

The data are plotted versus the dimensionless load parameters $W = w/mNLD(R/C)^2$ where w is the bearing load including the shaft weight and where:

m	=	oil viscosity, $lb/sec\text{-}in^2$
N	=	rotational speed, cps
L	=	bearing length, in
D	=	bearing diameter, in
R	=	bearing radius, in
C	=	bearing radial clearance, in

Use of the dimensionless load parameter W permits data taken at different speeds and at different oil film viscosity levels to be plotted against a single theoretical curve. In the tests performed here, the viscosity used in plotting the data was that corresponding to the bearing pad temperature as measured by the thermocouple imbedded just below the bearing babbitt 3/4 of the circumferential distance along the centerline of the pad.

The data presented in Figures 4 through 7 were corrected for the effect of the Hertzian deflection of the pivots supporting the pads. The effect of this Hertzian deflection is to change the BDC position of the bearing when load is applied relative to the BDC position measured when only the shaft weight was supported. This latter BDC position was obtained at the end of each ascending /descending load run by turning off the drive motor of the bearing test machine and measuring the position of the shaft as it came to rest under its own weight. Pivot deflection corrections for each load point were obtained by applying load to the non-rotating shaft. In this manner, it was determined that the effective pivot stiffness for each of the two bottom pads was 2.1×10^7 lb/in. The combined stiffness for vertical shaft motion was therefore:

$$2\,(2.1 \times 10^7)\,(\cos 45^o)^2 = 2.1 \times 10^7$$

The Hertz correction at the maximum applied load of 7000 lbs was only 0.00033 in.

As seen in Figures 4 and 6, the data obtained with the centrally pivoted bearing are in good agreement with theory throughout the load range for both the 1500 RPM and 3000 RPM runs. Data obtained with the 0.6 pivot bearing are in good agreement with theory at higher loads ($W > 12$ for 1500 RPM and $W > 6$ for 3000 RPM), but begin to deviate progressively from theory as the load W decreases below these values. Agreement between the vertical film thickness as measured by the x and y probes is generally very good except for the 0.6 pivot bearing at 3000 RPM where some disagreement between the x and y readings can be noted. A verification test was performed which demonstrated that this is due to a tighter assembled clearance between the x axis pads. This will be discussed in greater detail later. The theoretical curves plotted in Figures 4 through 7 predict that the film thickness obtained with a 0.6 pivot offset ratio should be slightly greater than that obtained with a centrally pivoted bearing, particularly at lighter loads. This predicted improvement in lightly loaded steady state performance, however, was not realized in the present tests

Dynamic Tests

Response to unbalance measurements were made for both the 0.5 and 0.6 pivot bearings at 7 values of applied load ranging from 1000 to 7000 lbs and at three different levels of rotating dynamic force. The rotating dynamic forces were produced by different combinations of unbalance weights attached to the ends of the test rotor. Figures 8 and 9 show typical plots of measured response amplitude vs applied load for both the 0.5 and 0.6 pivot bearing. Corresponding plots of measured phase angles are shown in Figures 10 and 11. Phase angle is defined as the angle by which the response lags the applied unbalance force, Data are plotted separately for the x probes and the y probes. Because of the isotropic nature of four pad, load between pad bearings, the responses measured by the x and y probes should be the same. Also, the responses occurring in the y plane measured by the y probes should be completely independent of the responses occurring in the x plane and vice versa. In a sense, then, the x and y probe measurements constitute two separate and theoretically independent tests of the dynamic characteristics of the bearing. Because of symmetry, the two "tests" should give redundant results. However, as can be seen in Figures 8 and 9, the amplitude of response as measured for by the x probes is significantly lower than that measured by the y probes, particularly in the case of the 0.6 pivot bearing. The discrepancy between the response characteristics measured in the x and y probe planes seen in Figures 8 through 11 was also observed in the tests made at the other two levels of dynamic loading (636 lbs and 1503 lbs).

It was speculated that the reduced response observed for the x probe readings could be due to a tighter than designed clearance between the x probe pads. To test this hypothesis, a dynamic test was made in which the upper bearing pads were removed. For this configuration, the x and y probe responses agreed quite closely with each other, supporting the above hypothesis. Lead wire checks of assembled clearances also confirmed that the x probe pads were assembled with a tighter clearance than was called for in the design.

Theoretical curves for predicted amplitude of response were calculated for the design clearance of 0.0055 inches and also for an assumed clearance of 0.004 inches. These curves are plotted in Figures 8 and 9. As can be seen, the "as designed" prediction agrees with the y probe data while the 0.004 inch clearance theoretical curve agrees well with the x probe data. This supports the contention that the discrepancy between the x and y probe results was due to a difference in the operating clearance.

Dynamic response characteristics measured at dynamic loadings of 686 pounds and 1503 pounds are shown in Figures 12 through 19. For simplicity, only the y probe data is shown, this being the data which conforms to the design clearance. According to linear dynamic theory, in which the bearing film dynamic properties are modeled by eight linear spring and damping coefficients, the amplitude of response should be linearly proportional to the dynamic force. The phase angle should be independent of the level of dynamic force.

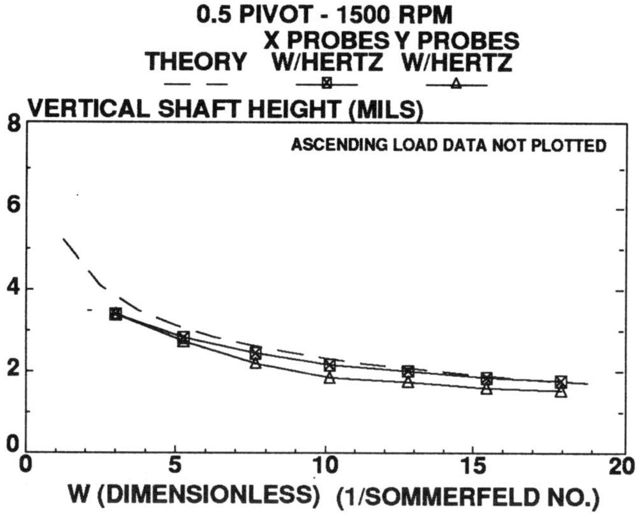

FIGURE 4 - BEARING FILM THICKNESS VS. LOAD

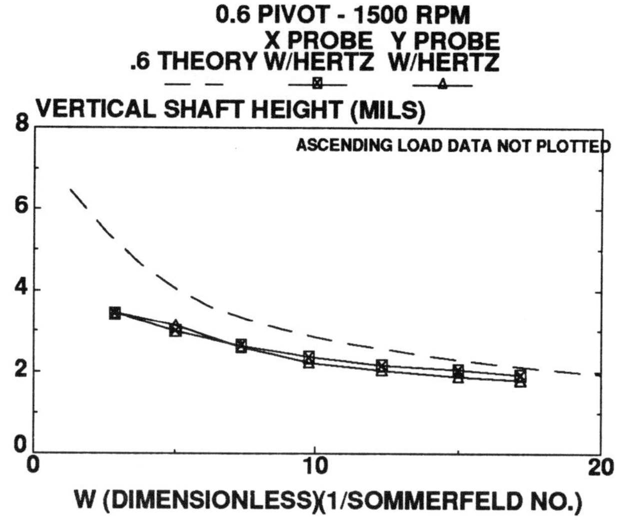

FIGURE 5 - BEARING FILM THICKNESS VS. LOAD

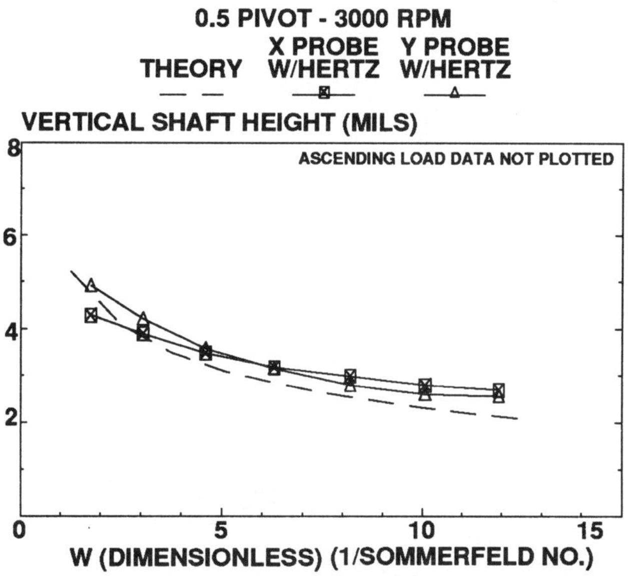

FIGURE 6 - BEARING FILM THICKNESS VS. LOAD

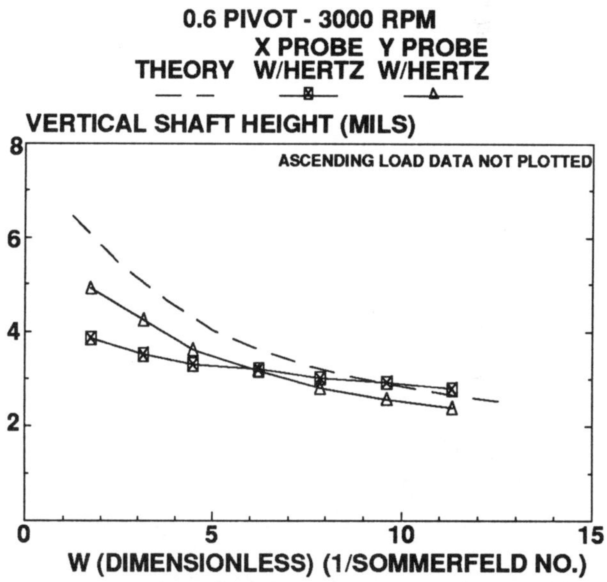

FIGURE 7 - BEARING FILM THICKNESS VS. LOAD

Looking at the plots shown in Figures 8 through 19, it is evident that the response behaviors were very linear over the whole range of the testing, which included response amplitudes up to 3 mils peak-to-peak. It is also noted that, in general, agreement between predicted and measured response characteristics of both the 0.5 pivot and 0.6 pivot bearings was very good. Particularly significant was the excellent agreement between measured and predicted phase angles for the tests conducted at the highest dynamic load where the response signal was strongest and, therefore, less subject to measurement error.

CONCLUSIONS

In general, the measured static and dynamic characteristics of both the 0.5 and 0.6 pivot test bearings were in good agreement with theoretical predictions. Particularly significant was the linear behavior of bearing response to different levels of dynamic loading.

The main objective of the testing was to confirm the enhanced bearing properties predicted for the 0.6 pivot bearing versus the 0.5 pivot bearing. The enhancements predicted were an increased film thickness and a higher damping coefficient.

The higher predicted film thickness was not obtained as evidenced by Figures 5 and 7. However the higher damping predicted for the 0.6 pivot bearing was achieved in the tests as determined by measurements of response phase angles. Response phase angle is a function of the ratio of bearing damping to bearing stiffness; the greater the damping, the higher the phase angle. As can be seen, most clearly in Figures 18 and 19, the predicted and measured phase angles for the 0.6 pivot bearing are greater than those for the 0.5 pivot bearing at applied loads above 3000 pounds. This was also true for the measurements made at dynamic loadings of 686 and 1172 pounds. Since our range of interest was above 5000 pounds, it was concluded that the 0.6 pivot bearing would offer improved dynamic performance for our application.

REFERENCES

1. Lund, J. W., "Spring and Damping Coefficients for the Tilting Pad Journal Bearing". Trans. ASLE, Vol. 7, 1964, pp. 342-352 .

ACKNOWLEDGMENT

The authors gratefully acknowledge the generosity of Kingsbury Incorporated in providing the bearings used in this test program.

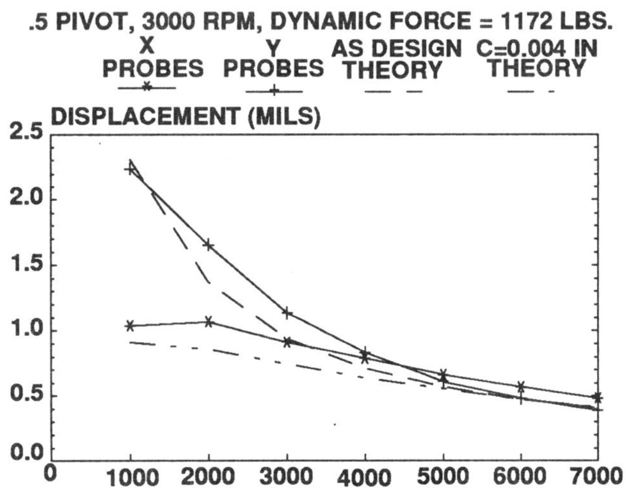

FIGURE 8 - DYNAMIC TESTS

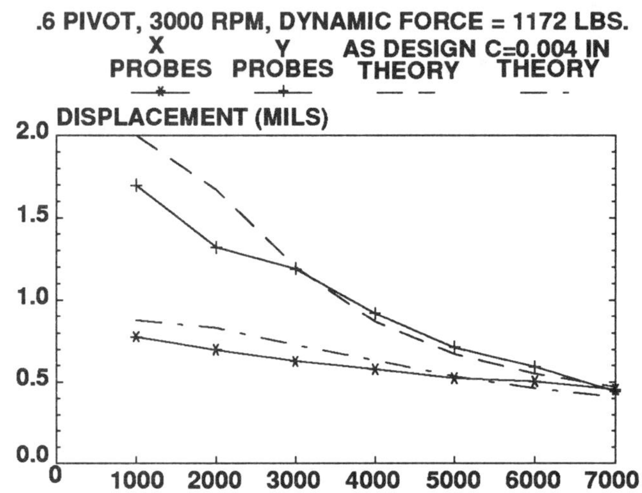

FIGURE 9 - DYNAMIC TESTS

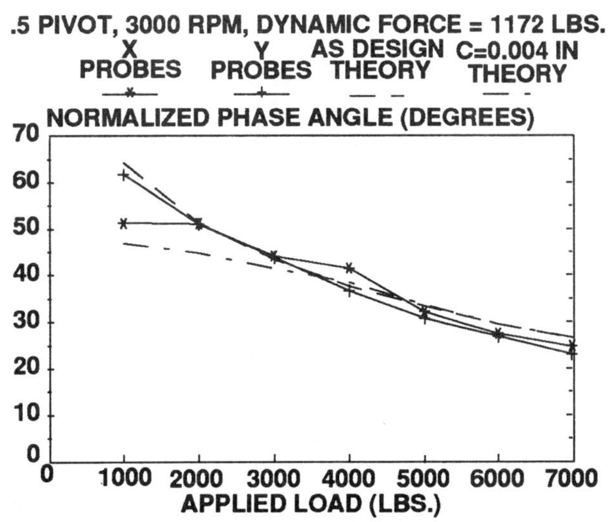

FIGURE 10 - DYNAMIC TESTS

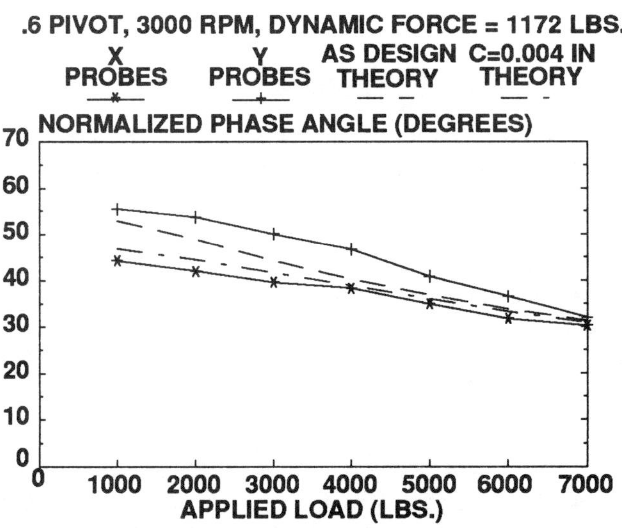

FIGURE 11 - DYNAMIC TESTS

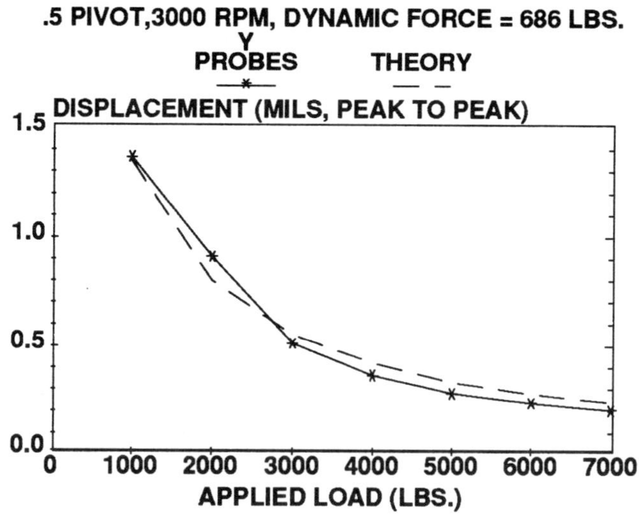

FIGURE 12 - DYNAMIC TESTS

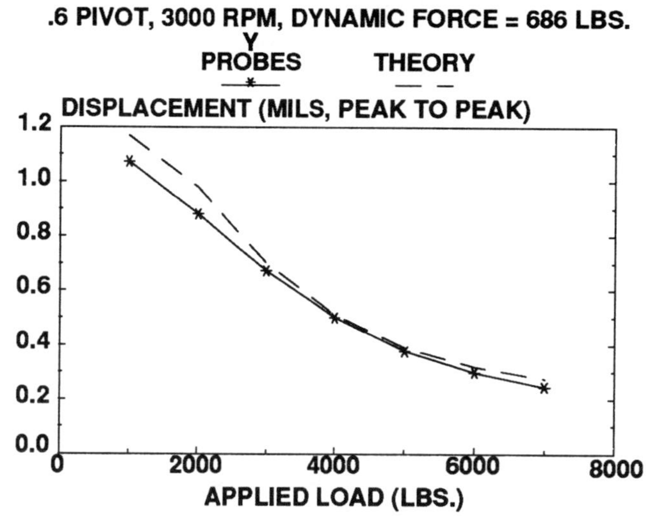

FIGURE 13 - DYNAMIC TESTS

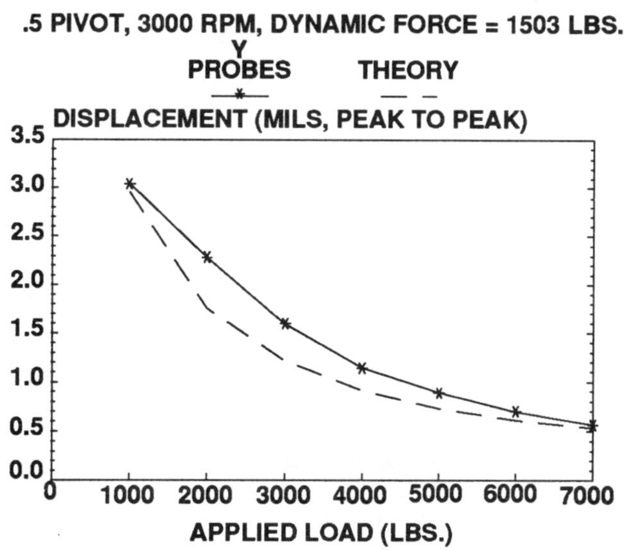

FIGURE 14 - DYNAMIC TESTS

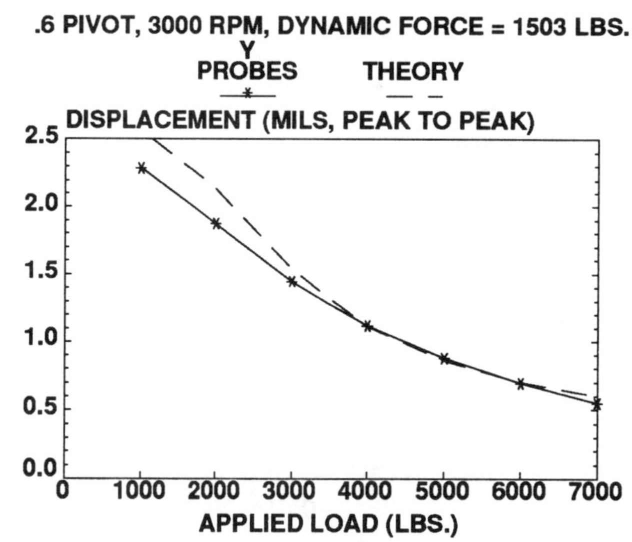

FIGURE 15 - DYNAMIC TESTS

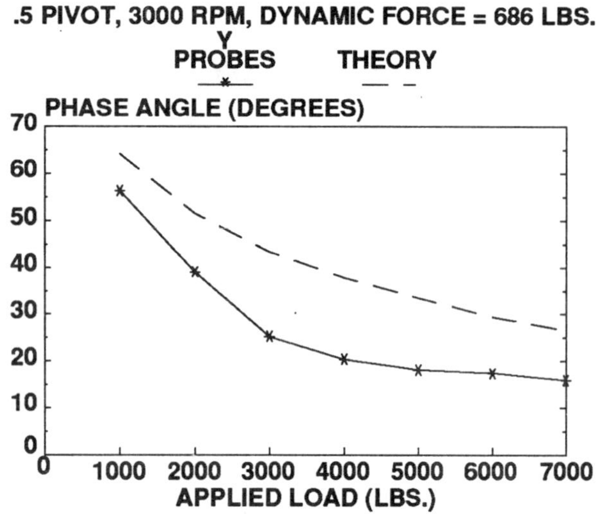

FIGURE 16 - DYNAMIC TESTS

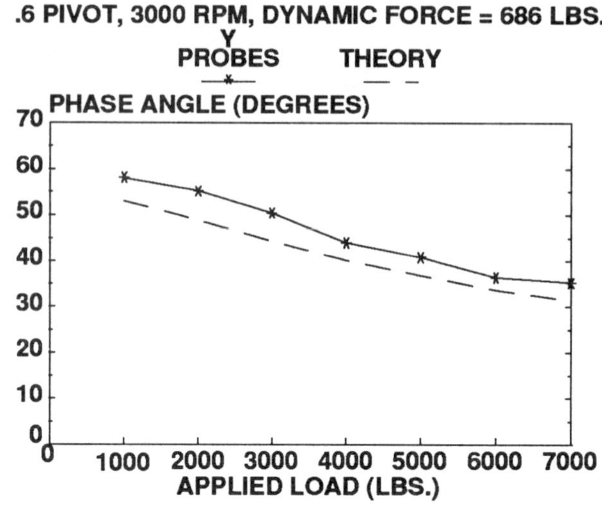

FIGURE 17 - DYNAMIC TESTS

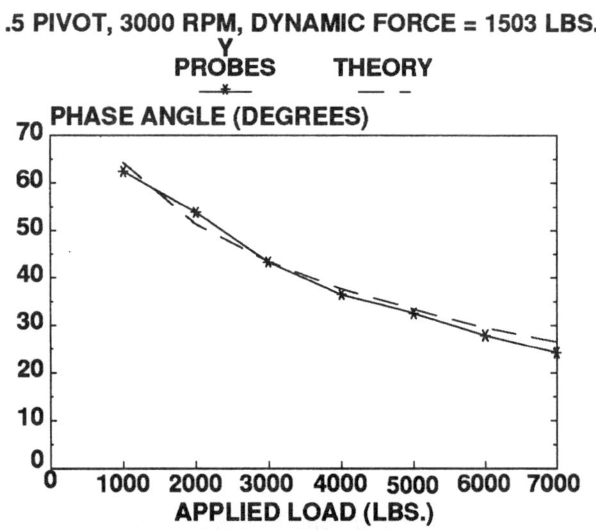

FIGURE 18 - DYNAMIC TESTS

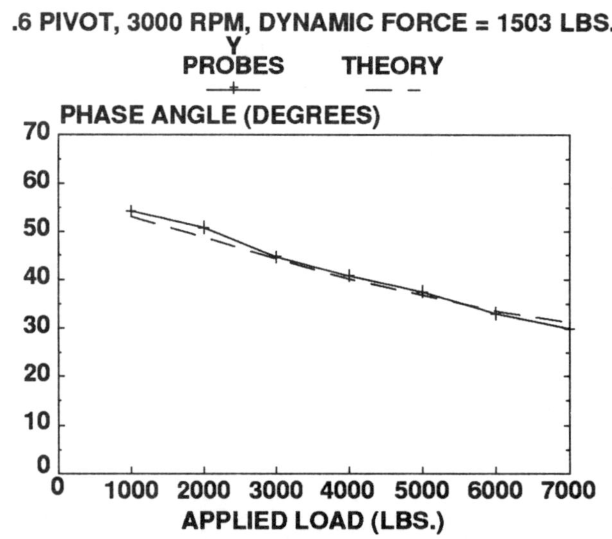

FIGURE 19 - DYNAMIC TESTS

DE-Vol. 63, Vibrations of Mechanical Systems
and the History of Mechanical Design
ASME 1993

DYNAMIC BEHAVIOR OF VEHICLES TRAVELLING ON BRIDGES

Ebrahim Esmailzadeh and Mehrdaad Ghorashi
Department of Mechanical Engineering
Sharif University of Technology
Tehran, Iran

ABSTRACT

An investigation into the dynamic behavior of a bridge with simply supported boundary conditions, carrying a moving vehicle, is performed. The vehicle has been modelled as a two degree of freedom lumped-parameter system travelling at a uniform speed. Furthermore, the bridge is assumed to obey the Euler-Bernoulli beam theory of vibration. This analysis may well be applied to a beam with different boundary conditions, but the computer simulation results given in this paper are set for only the case of freely hinged ends. Numerical solutions for the derived differential equations of motion are obtained and their close agreement, in some extreme cases, with those reported earlier by the authors are observed. Finally, the effect of speed on the maximum dynamic deflection of bridge is shown to be of much importance and hence an estimation for the critical speed of the vehicle is presented.

INTRODUCTION

The analysis of structures carrying moving loads is of considerable practical importance. Bridges on which vehicles or trains travel, and trolleys of overhead travelling cranes that move on their girders may be modelled as moving masses traversing along a simply supported beam.

Since the middle of the last century, when railway construction began, the problem of oscillation of bridges under travelling loads has interested many engineers. Contributions to the solution of this problem were made by Stokes (1849), Willis, et al. (1849), and many others. Timoshenko (1927) considered the problem of a pulsating load passing over a bridge. Inglis (1934), in his systematic analysis of trains crossing a bridge, took into account many important factors such as the effects of moving loads, the influence of damping, and the spring suspension of the locomotive.

For the case of a concentrated force moving with a constant velocity along a beam, neglecting damping forces, Timoshenko (1974) found a solution and presented an expression for the critical velocity. Stanisic and Hardin (1969), tried to solve the dynamic behavior of a simply supported beam carrying a moving mass, which is interesting enough, but not easily applicable to different boundary conditions. A comprehensive treatment on the subject of vibrations of structures due to moving loads, which contains a large number of related cases, is that of Fryba (1972).

In an earlier work Esmailzadeh, and Ghorashi (1992) investigated the behavior of a beam carrying a moving concentrated mass, where in, the inertial effect of the mass was taken into account. As an extension to that work, the present paper deals with the problem of vibration analysis of bridges subjected to a mobil two degree of freedom vehicle model. It is expected that considering resilience of the vehicle suspension system leads to a more realistic simulation of the earlier models. Clearly, the rigid point mass formulation can be considered as a special case of the present analysis, if the suspension system is eliminated altogether.

In this paper the following assumptions were made:

1- The bridge characteristic is described by the Euler-Bernoulli beam equation, since it is of a constant cross section with a uniform mass distribution.

2- The vehicle travels at a constant speed with its wheels keep contact with the bridge surface at all times.

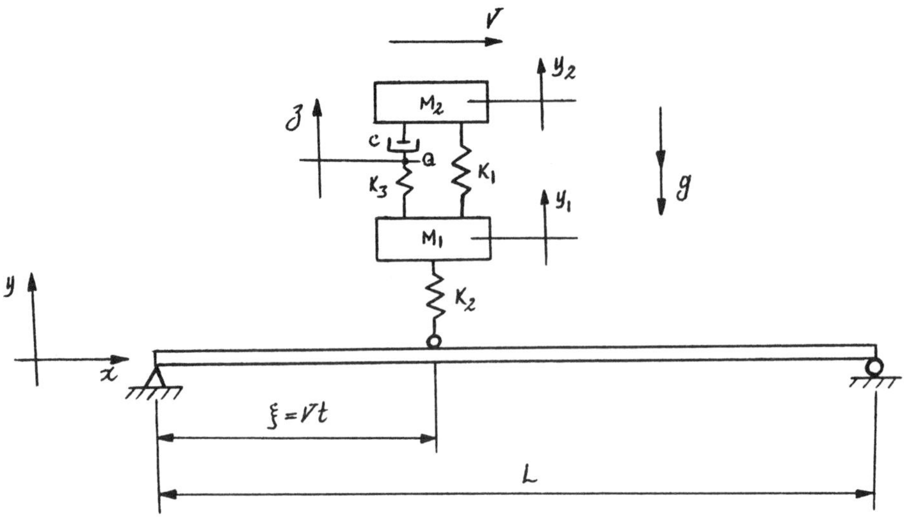

Figure 1 The mathematical model of the problem.

3- The computations are performed for the case of a beam with a simply supported boundary conditions.

4- The bridge is assumed to carry the vehicle, in its initial state, at its left-hand support, as illustrated in Figure 1.

ANALYTICAL FORMULATION

Consider the case of a vehicle advancing along a bridge with a constant velocity ***V***. The vehicle is modelled as a two degree of freedom system as shown in Figure 1. The system has a sprung mass M_2 being supported by the suspension system in the form of the main spring K_1, and the visco-elastic damper of bushing K_3, with the damping coefficient ***C***. The vehicle tire stiffness is denoted by K_2 and the unsprung mass, M_1. All these elements are assumed to work in their linear range.

It must be pointed out that although the supporting beam is considered as simply supported but the formulation is general for any kind of boundary conditions.

According to the Euler-Bernoulli beam theory, the governing differential equation describing the lateral vibration of a beam, carrying the time varying force, ***f(x,t)*** per unit length is (Tse, et al., 1978):

$$EI\frac{\partial^4 y}{\partial x^4}+m\frac{\partial^2 y}{\partial t^2}=f(x,t) \qquad \textbf{(1)}$$

where: E=modulus of elasticity; I=second moment of area of the beam's cross section; m=mass per unit length of the beam, and ***y(x,t)*** the deflection of the beam, being measured from its equilibrium position when unloaded. Hence, ***f(x,t)*** can be expressed as:

$$f(x,t)=-[(M_1+M_2)g+K_2[y(\xi,t)-y_1(t)]]\delta(x-\xi) \qquad \textbf{(2)}$$

The differential equations of motion for the body, the wheel, and the connecting point, Q are:

$$M_2\frac{d^2y_2}{dt^2}=-K_1(y_2-y_1)-C\left(\frac{dy_2}{dt}-\frac{dz}{dt}\right) \qquad \textbf{(3)}$$

$$M_1\frac{d^2y_1}{dt^2}=K_1(y_2-y_1)+K_3(z-y_1)-K_2(y_1-y) \qquad \textbf{(4)}$$

$$C\left(\frac{dy_2}{dt}-\frac{dz}{dt}\right)-K_3(z-y_1)=0 \qquad \textbf{(5)}$$

In order to solve the set of five differential equations, i.e., (1) to (5), the transverse vibration of the beam may be assumed to be:

$$y(x,t)=\sum_{i=1}^{\infty}\phi_i(t)X_i(x) \qquad \textbf{(6)}$$

where ϕ_i 's are unknown functions of time and

X_i's are the normalized deflection curves for the ***i***-th mode of the vibrating beam. Substituting (6) in (1), the governing equation of beam vibrations, as shown in the appendix, is found to be:

$$EI\sum_{i=1}^{\infty}X_i^{iv}(x)\phi_i(t)+m\sum_{i=1}^{\infty}X_i(x)\phi_i''(t)= -\left[(M_1+M_2)g+K_2\sum_{j=1}^{\infty}X_j(\xi)\phi_j(t)-K_2y_1(t)\right]\times \sum_{i=1}^{\infty}X_i(\xi)X_i(x) \qquad \textbf{(7)}$$

Recalling the following equation for the free vibration of an Euler-Bernoulli beam,

$$X_i^{iv}(x)-\beta_i^4 X_i=0 \qquad (8)$$

where: $\beta_i^4=\frac{mp_i^2}{EI}$, and p_i is the i-th natural frequency of the beam, equation (7) would then be reduced to:

$$\left[(M_1+M_2)g+K_2\sum_{j=1}^{\infty}X_j(vt)\phi_j(t)-K_2y_1(t)\right]X_i(vt)+ m\frac{d^2\phi_i(t)}{dt^2}+mp_i^2\phi_i(t)=0 \quad ; \quad i=1,2,3,... \qquad (9)$$

If N is the number of modal shapes to be considered in the solution, the equations (3),(4),(5) and (9) would then form a set of ***(N+3)*** unknowns,viz., $y_1(t)$, $y_2(t)$, $z(t)$ and $\phi_i(t)$ $(i=1,2,3,...,N)$ to be evaluated at any time in the interval of zero and ***L/V***.

NUMERICAL SOLUTIONS

To solve these equations numerically, approximate central difference formulae have been utilized for the derivatives (Mathews, 1987). Hence, the above differential equations are transformed to a set of linear algebraic equations, which would then be solved for each interval of time. It can be shown that at an arbitrary instant of time, t_k , the following equations hold true,

$$\phi_i(t_{k+1})=2\phi_i(t_k)-\phi_i(t_{k-1})-\frac{h^2}{m}\times [K_2X_i(vt_k)\times\sum_{j=1}^{N}X_j(vt_k)\phi_j(t_k)]- \frac{h^2}{m}[EI\beta_i^4\phi_i(t_k)-K_2X_i(vt_k)y_1(t_k)+ (M_1+M_2)gX_i(vt_k)] \quad i=1,2,3,...,N \qquad (10)$$

$$y_1(t_{k+1})=2y_1(t_k)-y_1(t_{k-1})+\frac{h^2}{M_1}[K_2\sum_{i=1}^{N}X_i(vt_k)\times \phi_i(t_k)]+\frac{h^2}{M_1}[-(K_1+K_2+K_3)y_1(t_k)+K_1y_2(t_k)+ K_3z(t_k)] \qquad (11)$$

$$y_2(t_{k+1})=2y_2(t_k)-y_2(t_{k-1})+\frac{h^2}{M_2}\times [K_1(y_1(t_k)-y_2(t_k))-K_3(z(t_k)-y_1(t_k))] \qquad (12)$$

$$z(t_{k+1})=z(t_{k-1})+2y_2(t_k)-2y_2(t_{k-1})-\frac{2hK_3}{C}[z(t_k)- y_1(t_k)]+\frac{h^2}{M_2}[K_1(y_1(t_k)-y_2(t_k))-K_3(z(t_k)-y_1(t_k))] \qquad (13)$$

where k being the current time step number, and h is the time increment. Equations (10) through (13) could be used to calculate y_1, y_2, z, and ϕ_i 's at any time during the traverse of vehicle on the bridge. In order to start the process, initial and boundary conditions should be implemented. Considering the simply supported beam, the normalized deflection functions are:

$$X_i(x)=\sqrt{\frac{2}{L}}\sin(\frac{i\pi x}{L})\,;\, i=1,2,3,... \qquad (14)$$

Furthermore, regarding the rest position at initial state, for small values of h one can deduce:

$$y_1(t_0)=y_1(t_1)=y_2(t_0)=y_2(t_1)=0 \quad z(t_0)=z(t_1)=\phi_i(t_0)=0 \quad \phi_i(t_1)=0\ ;\ i=1,2,3,... \qquad (15)$$

Substituting (14) and (15) into equations (10) through (13) would lead to a set of equations readily applicable to any set of illustrative data. Regarding the approximations involved, in order to ensure the stability and convergence of the solution, smaller time steps must be utilized.

A computer program is developed on the basis of the above mentioned numerical technique which is a further extension to that reported by the authors (1992). Initially, for the sake of comparisons, it has been run with the following numerical data which are the same as those used by Esmailzadeh, and Ghorashi, (1992).

$$E=2.07\times10^{11}\ N/m^2;\ I=1.04\times10^{-6}\ m^4;\ V=12\ Km/h \quad a=174.9\ m^2/s;\ M=70\ Kg;\ g=9.81\ m/s^2 \quad m=7.04\ Kg/m;\ h=0.01\ s;\ L=10\ m \qquad (16)$$

Figure 2 illustrates the dynamic response of the considered beam when the suspension system is assumed to behave rigidly. In comparing this with the corresponding plots of Esmailzadeh, and Ghorashi, it can be shown that the formulations derived for the present system is a general case for the rigid mass model.

Using the following set of data, the dynamic behavior of vehicle and bridge have been determined.

$$E=2.07\times10^{11}\ N/m^2;\ I=0.87\ m^4 \quad V=10\ m/s;\ a=3000\ m^2/s \quad M_1=600\ Kg;\ M_2=3000\ Kg \quad m=20000\ Kg/m;\ L=100\ m \quad g=9.81\ m/s^2;\ C=100\ Ns/m \quad K_1=5000\ N/m;\ K_2=2000\ N/m \quad K_3=500\ N/m;\ h=0.025\ s \qquad (17)$$

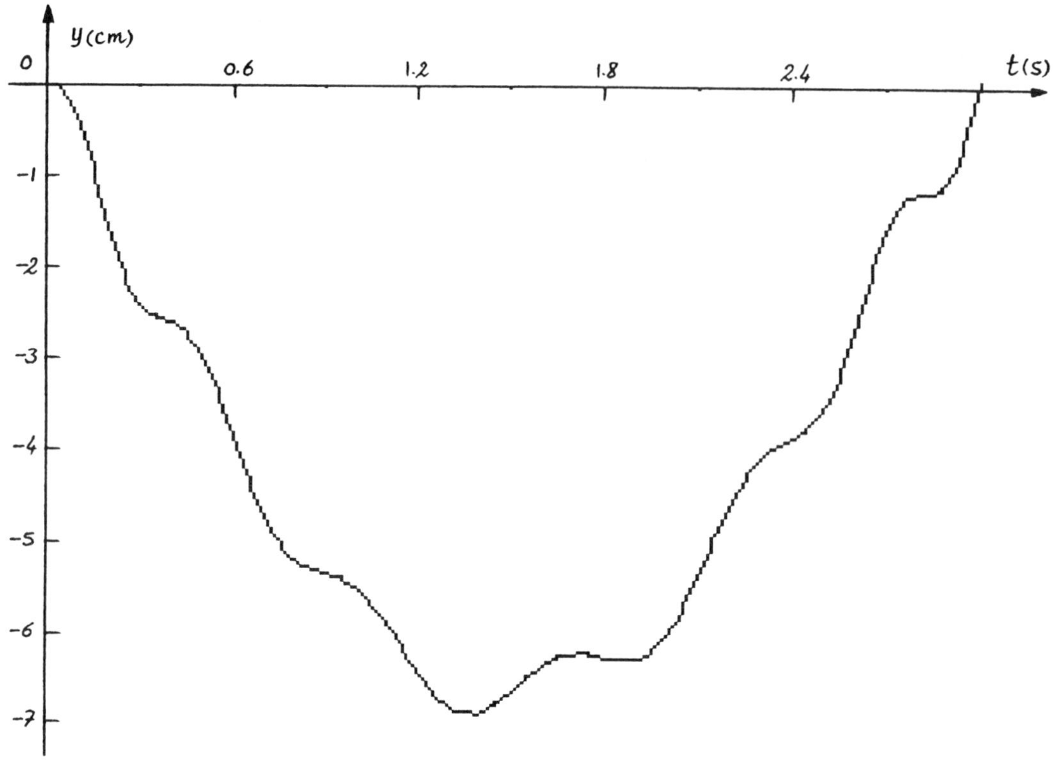

Figure 2 Time history diagram of the deflection at the mid-point of the beam. The vehicle suspension has been assumed to be rigid with the data as given by (16).

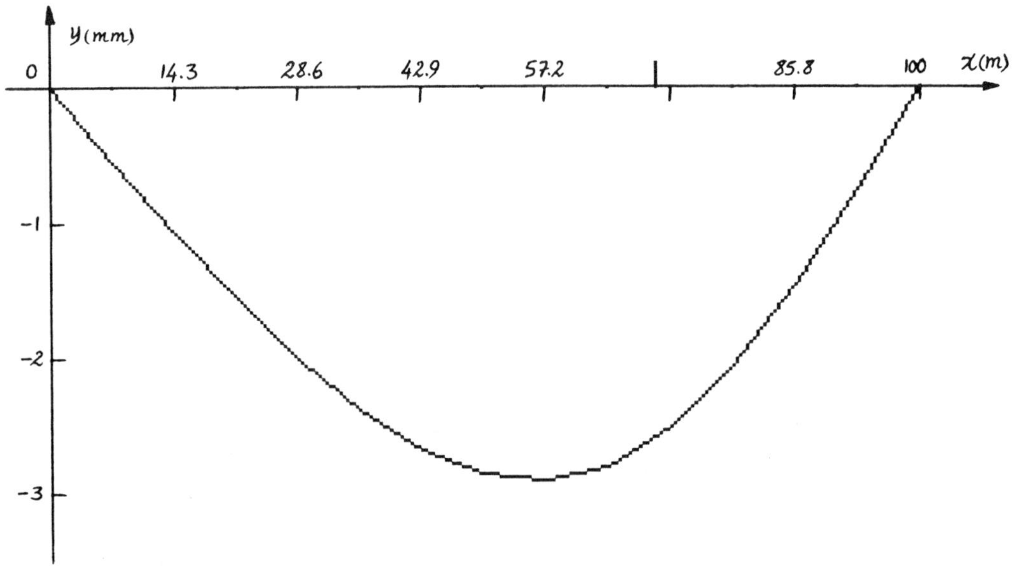

Figure 3 Bridge deflected shape; t=7 s, data as given by (17). The instantaneous position of the vehicle is also shown.

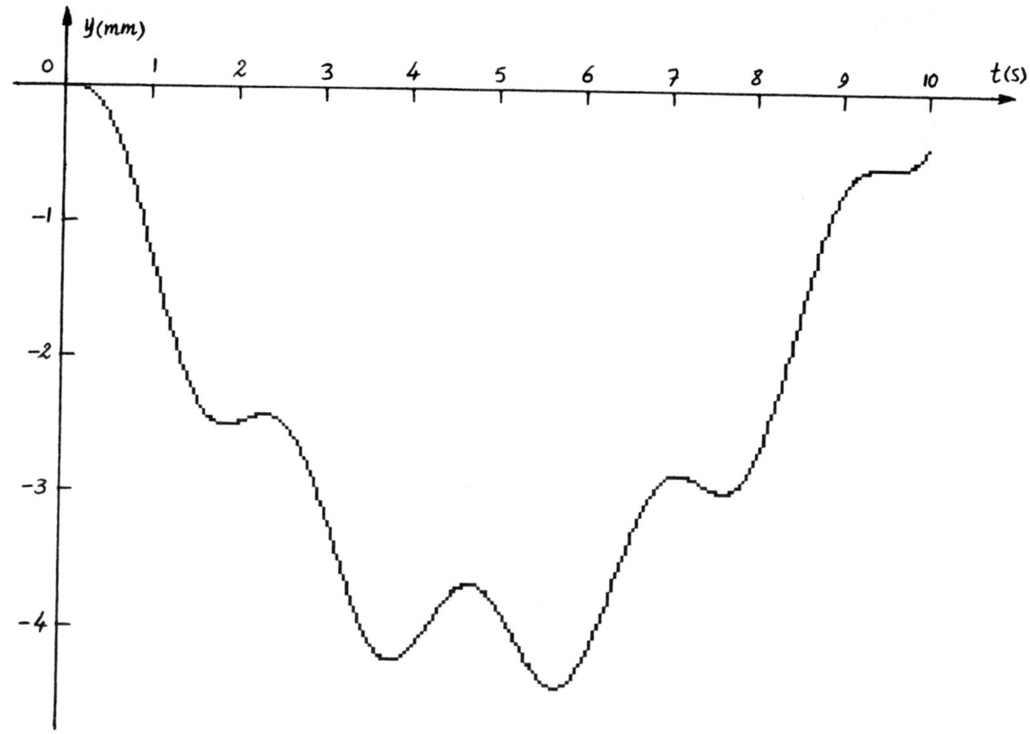

Figure 4 Time-history diagram of the deflection at the mid-point of the bridge.

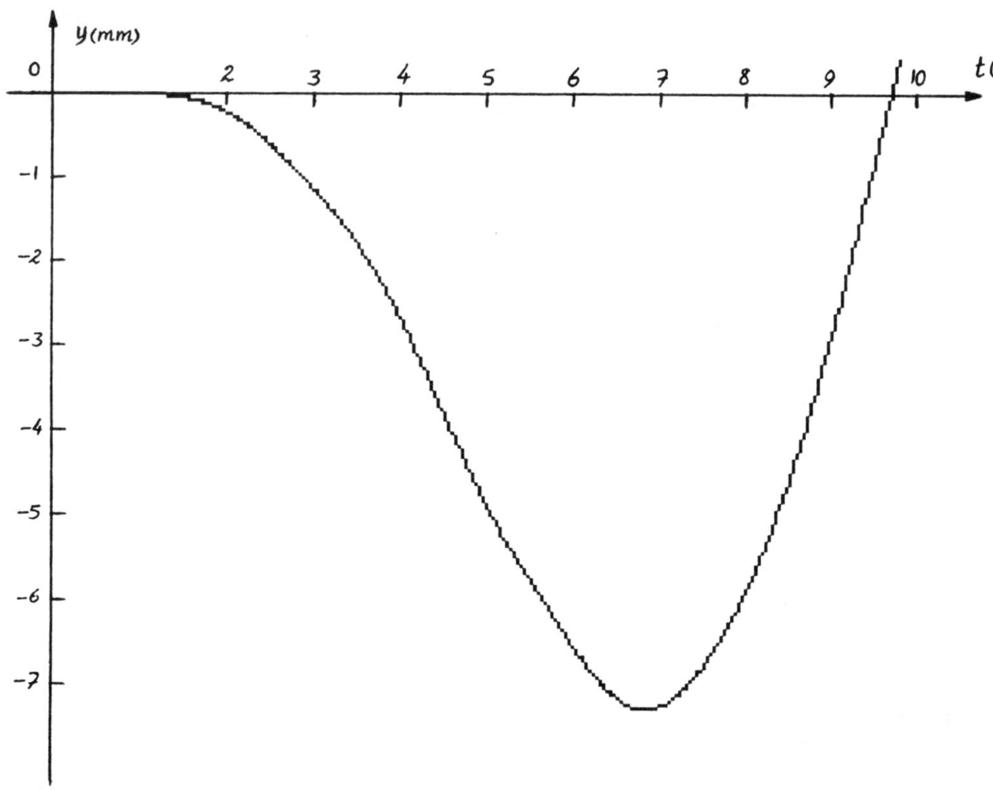

Figure 5 Time history diagram of the vehicle model sprung mass.

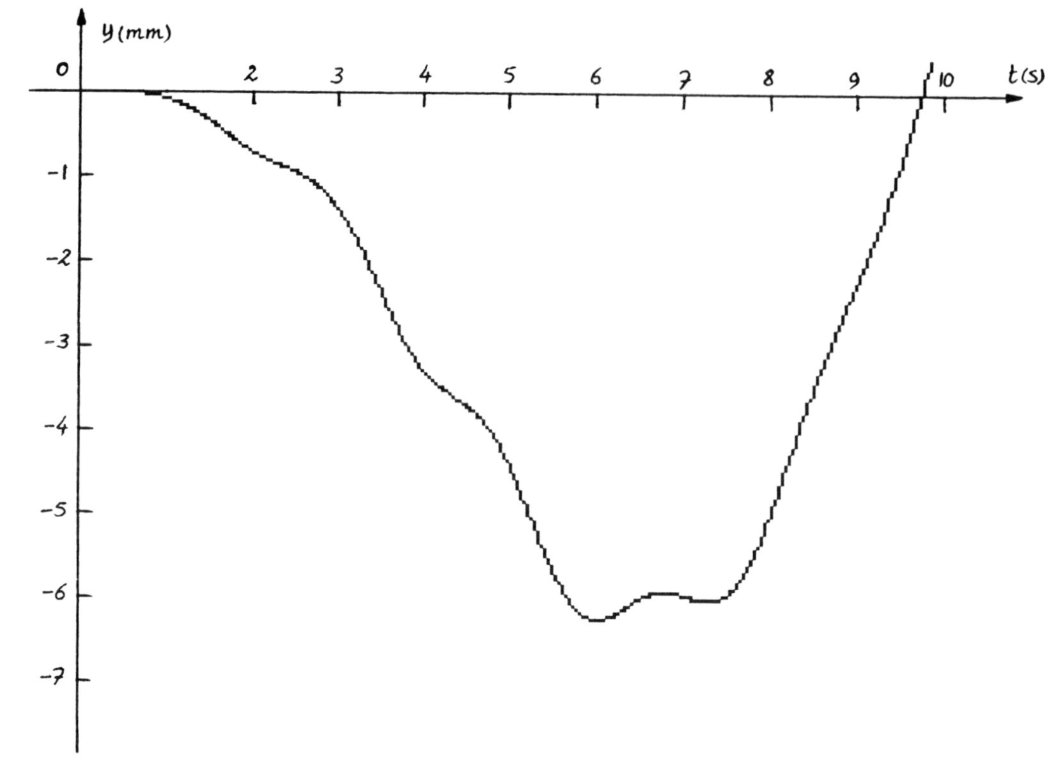

Figure 6 Time history diagram of the vehicle model unsprung mass.

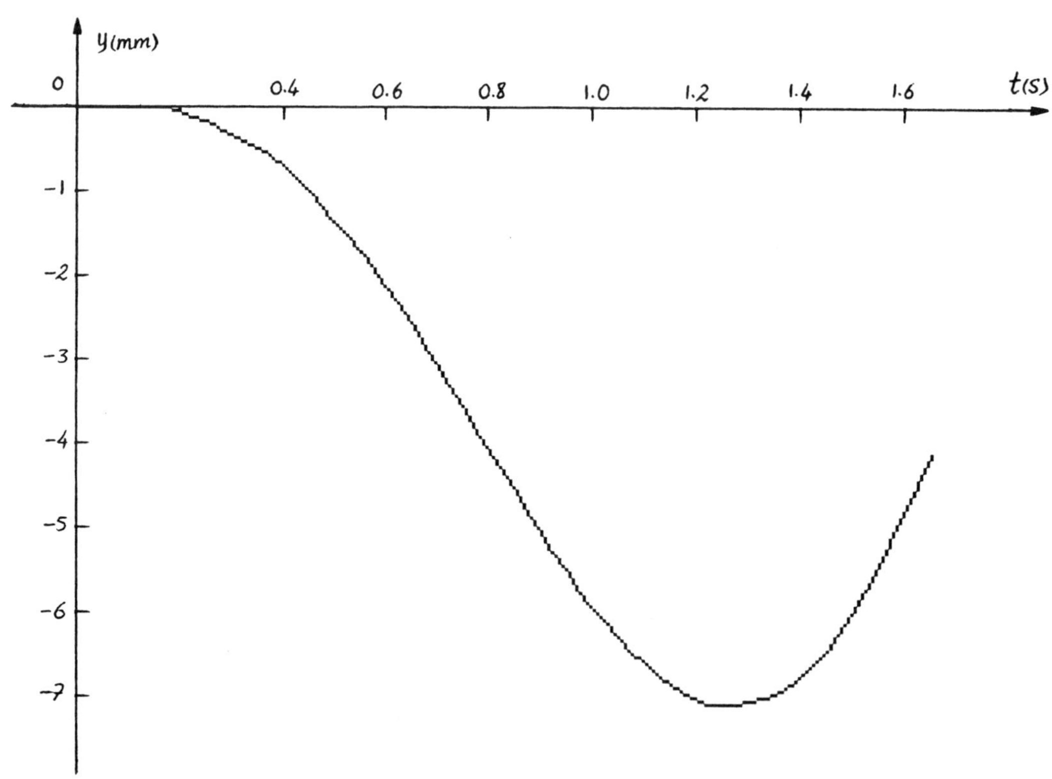

Figure 7 Variation of the deflection at the middle of the bridge, at the critical speed.

Table I- Variation of the maximum dynamic deflection of bridge, D_m, with the velocity of the vehicle.

V	m/s	10	20	30	40	50	60	70
D_m	mm	4.4	4.6	6.0	6.7	7.0	7.1	7.0

The elastic deflection of bridge for an arbitrary chosen time is shown in Figure 3. Moreover, Figures 4, 5, and 6 illustrate the time history diagrams for the mid-point of bridge, vehicle sprung mass, and vehicle unsprung mass, respectively. All these computations have been performed for the first four modes of vibration. It is observed that the number of modal shapes considered in the analysis, could only slightly affect the results. This would in turn ensure the rapid convergence of the numerical solution.

Finally, considering the maximum deflection at the mid-point of the beam, the critical value of the vehicle speed has been calculated. According to Table 1, the value of the critical speed for the moving vehicle model is about 60 *m/s*. The time history diagram at the mid-span of the bridge with the vehicle travelling at critical speed has been illustrated in Figure 7.

CONCLUSION

The dynamic performance of a bridge, with simply supported end conditions, when carrying a two degree of freedom vehicle model which moves at a uniform speed has been analyzed. In order to solve the governing differential equation, the technique of central difference expansions was employed. The solution series converges rapidly and usually requires a few number of modal shapes for obtaining relatively accurate results. Considering the variation of maximum dynamic deflection of bridge with vehicle speed, the critical value for this speed was determined. Furthermore, it was observed that in the special case of the system with rigid suspension, representing a concentrated moving mass, the results obtained are in close agreement with those reported earlier by the authors.

REFERENCES

Esmailzadeh, E., and Ghorashi, M., 1992, "Vibration Analysis of Beams Traversed by Moving Masses," *Proceedings of the Int. Conf. on Eng. Appl. of Mech.*, Tehran, Iran, Vol. 2, pp. 232-238.

Fryba, L., 1972, *Vibration of Solids and Structures Under Moving Loads*, Noordhoff International Publishing, Groningen.

Inglis, C.E., 1934, *A Mathematical Treatise on Vibration in Railway Bridges*, The University Press, Cambridge.

Mathews, J.H., 1987, *Numerical Methods*, Prentice-Hall Inc.

Stanisic, M.M., and Hardin, J.C., 1969, "On Response of Beams to an Arbitrary Number of Moving Masses," *J. of Franklin Inst.*, Vol. 287, pp. 115.

Stokes, G.G., 1849, "Discussion of a Differential Equation Relating to the Breaking of Railway Bridges," *Trans. Cambridge Philosoph. Soc.*, 8, Part 5, pp. 707-735.

Timoshenko, S., 1927, "Vibration of Bridges," *Trans. ASME*, Vol. 49-50, Part II, pp. 53-61.

Timoshenko, S., Young, D.H., and Weaver, W., 1974, *Vibration Problems in Engineering*, 4th Ed., John Wiley & Sons, New York.

Tse, F.S., Morse, I.E., and Hinkle, R.T., 1978, *Mechanical Vibrations, Theory and Applications*, 2nd Ed. Allyn and Bacon, Inc., Boston.

Willis, R., et al., 1849, Appendix to: *Report of the Commissioners Appointed to Inquire into the Application of Iron to Railway Structures*, W. Clowes and Sons, London.

APPENDIX

In order to derive equation (7), the function ***f(x,t)*** is assumed to be expressible as in the following series:

$$f(x,t)=\sum_{i=1}^{\infty} X_i(x)\,\psi_i(t) \qquad \textbf{(A1)}$$

where $\psi_i(t)'s$ are unknown functions of time.

Substituting for $y(\xi,t)$ from equation (6), and multiplying both sides of the resulting equation by $X_j(x)$, and integrating along the length of the bridge, the functions $\psi_i(t)$ are obtained as:

$$\psi_i(t)=-[(M_1+M_2)g+K_2\sum_{j=1}^{\infty} X_j(\xi)\phi_j(t)-K_2y_1(t)]X_i(\xi) \quad ; \quad i=1,2,3,... \qquad \textbf{(A2)}$$

Equation (7) is, therefore, obtained by substituting from equations (6) and (A2) in equation (1).

DE-Vol. 63, Vibrations of Mechanical Systems
and the History of Mechanical Design
ASME 1993

INELASTIC ANALYSIS OF REINFORCED CONCRETE SHEAR WALL STRUCTURES UNDER SEISMIC EXCITATION

Ming L. Wang
Department of Civil Engineering
The University of New Mexico
Albuquerque, New Mexico

ABSTRACT

During strong ground motions, members of reinforced concrete structures undergo cyclic deformations and experience permanent damage. Members may lose their initial stiffness as well as strength. Recently, Los Alamos National Laboratory has performed experiments on scale models of shear wall structures subjected to recorded earthquake signals. In general, the results indicated that the measured structural stiffness decreased with increased levels of excitation in the linear response region. Furthermore, a significant reduction in strength as well as in stiffness was also observed in the inelastic range. Since the in-structure floor response spectra, which are used to design and qualify safety equipment, have been based on calculated structural stiffness and frequencies, it is possible that certain safety equipment could experience greater seismic loads than specified for qualification due to stiffness reduction. In this research, a hysteresis model based on the concept of accumulated damage has been developed to account for this stiffness degradation both in the linear and inelastic ranges. Single and three degrees of freedom seismic Category I structures were analyzed and compared with equivalent linear stiffness degradation models in terms of maximum displacement responses, permanent displacement, and floor response spectra. The results indicate significant differences in responses between the hysteresis model and equivalent linear stiffness degradation models. The hysteresis model is recommended in the analysis of reinforced concrete shear-wall structures to obtain the in-structure floor response spectra for equipment qualification. Results of both cumulative and one shot tests are compared.

INTRODUCTION

In recent Los Alamos National Laboratory test programs, both free vibration and small amplitude excitation tests have been performed, and scale models of shear wall structures were subjected to a set of recorded earthquake time history simulations (the 1940 N-S El Centro corrected data). The amplitude of the simulated earthquake time history was increased in steps for successive tests, to assess the effects of peak ground acceleration on response. The tests indicated that a significant difference existed between calculated shear wall stiffnesses and those inferred from both static and dynamic tests; measured stiffnesses were found to be 70 to 80 percent of the computed values. During the simulated earthquake time history tests, measured structure stiffnesses were found to decrease with increased levels of excitation, and reduction in stiffnesses of up to 25 percent of the calculated stiffness values were measured, which corresponds to a reduction in structural frequencies of up to 50 percent [1-6].

The observed differences between calculated and measured stiffnesses and frequencies represent a potentially important issue related to the seismic design and safety of nuclear power plants. Most Category I structures at typical PWR and BWR power plants have calculated fixed base frequencies in the 5 Hz to 20 Hz range. Based on the LANL tests, and depending on the level and frequency content of the earthquake time history and the local soil conditions, these structures could have effective frequency reductions of up to 50 percent, which is in the range of 2.5 Hz to 10 Hz. Also, most broad band strong motion recorded earthquake accelerograms have the majority of their energy in the 2 Hz to 8 Hz range. Thus if the structure has an effective frequency in this band, the excitation of the structure would be greater than was considered in the design of the structure. Since the design in-structure floor response spectra which are used to design and qualify safety equipment have been based on calculated structural stiffnesses and frequencies, it is possible that certain safety equipment could experience greater seismic loads than specified for qualification. In general, the cyclic degradation of stiffness was observed in elements of both flexure and shear deformation. A significant reduction in strength as well as in stiffness was also observed after cycling the element in the inelastic range. A hysteresis model based on the concept of accumulated damage was developed to account for both stiffness and strength reductions. Single-degree-of-freedom (SDF) and multi-degree-of-freedom (MDF) shear wall structures were analyzed. The results are compared with linear analysis without considering the stiffness degradation and an equivalent linear stiffness degradation analysis in terms of maximum displacement responses, permanent displacement, maximum shear stress and floor response spectra.

The results indicate significant differences in response between hysteresis models and equivalent linear stiffness degradation and linear models. Equivalent stiffness degradation models tended to soften the structure and underpredict the floor acceleration responses. On the other hand, the hysteresis models predicted a wider band of frequency distribution and a larger magnitude of absolute acceleration responses. Hysteresis models can be used in the analysis to obtain the in-structure floor response spectra for equipment qualification. Ductility and hysteresis energy dissipation are significant in the analysis of reinforced concrete shear wall structures.

Cumulative Damage Concept

The damage to a reinforced concrete member or an assemblage subjected to cyclic loading is assumed to depend on the maximum displacements experienced during each cycle. No damage is accumulated before yielding. This is consistent with previous experiments (Farrar, Dove, & Baker, 1990); however, some experiments show that the stiffness of a member may continue to decrease when subjected repeatedly to the same maximum displacement after yielding (Townsend & Hanson, 1977). This stiffness decrease is shown in Figure 1. The cyclic-loading parameter β is defined as:

$$\beta = c \sum_{i=1}^{N} \frac{X_i}{X_f} \tag{1}$$

where X_i is the maximum displacement in i^{th} cycle, X_f is failure displacement under monotonically increasing loading, and c is a constant less than one. Note that only the displacements occurring in one direction are accumulated. For unsymmetrical loading of cross-sections, a separate value of b for each direction must be used.

A non-dimensional and non-negative scalar quantity D is introduced and is termed in a damage parameter whose value varies between 0 and 1. The damage parameter is a function of loading parameter β such that

$$D = f(\beta); \; f(0) = 0; \; f(1) = 1. \tag{2}$$

Note that D is a state variable which changes with b. Assuming for simplicity that the rate of change of damage $dD/d\beta$ varies linearly with D, the following damage accumulation relation can be developed:

$$\frac{dD}{d\beta} = nD \tag{3}$$

where n is a constant to be determined from experimental data. The solution of the differential equation (3) is:

$$D = f(\beta) = \frac{e^{n\beta} - 1}{e^{n} - 1}. \tag{4}$$

Analytical Model

It is assumed that a single-degree-of-freedom reinforced concrete structure or component has a unique monotonic load-displacement curve in each direction of loading before any damage has occurred. For simplicity, a bilinear skeleton curve is assumed (Fig. 2). Note that Fy and Xy are the load and displacement corresponding to the yielding point, $Ky = Fy/Xy$ is initial stiffness and aKy is the post-yield stiffness.

The damage due to cyclic loading is expressed in terms of the decrease in the yielding-load carrying capacity. The reduction in the yield strength (from point B to G in Fig. 2) can be written as:

$$F_D(X_i) = F(X_i)\,(1-D) \tag{5}$$

where X_i is the maximum displacement the structure has experienced during the ith cycle, and $F(X_i)$ is the load corresponding to the displacement X_i on the skeleton load-displacement curve. Thus the yield-load carrying capacity of the structure is reduced from $F(X_i)$ to $F_D(X_i)$, as shown in Figure 2, after it has experienced damage D which will depend on the cyclic loading parameter β. Based on Equations 1, 4, and 5, a hysteresis model that can account for stiffness degradation and yield-strength deterioration is proposed. The procedure for a symmetrical loading and a symmetrical cross section is outlined below.

1. The model operates on a bilinear skeleton curve defined by Fy, Ky and α. Once the section has yielded, unloading is associated with the smallest stiffness observed in the same loading direction. Before the completion of the first cycle the smallest stiffness observed is equal to Ky. After that, the smallest stiffness is defined by $K(X_i)$, which is the line connecting points G and C in Figure 2. The point G is determined using Equation 5, and for subsequent loading, it is the current value of the yield point whereas the point C is determined from X_i and Ky.
2. When the load is reversed, the stiffness is determined by considering the slope of the line connecting the load reversing point (points C or F) to the current value of the yield point (points D or G).
3. When the loading is continued beyond the maximum displacement previously experienced in that direction (such as X_i), the loading displacement response is determined by assuming the same post-yield stiffness as that for the skeleton curve (line GH in Fig. 2).

4. If during any cycle the maximum displacement is less than the current yielding point, no damage is accumulated (point F ' in Fig. 2).

The proposed modal requires the following quantities: Fy, Ky and a, needed for evaluating the skeleton curve, and the quantities X_F, C and n to define the relative extent of damage. The parameters needed for the proposed hysteresis model were obtained from the results of a small scale model of a reinforced concrete beam-column joint. The dimensions of the specimen are shown in Figure 3. The specimen was designed according to the seismic design requirement of ACI 318-88 and the recommendation of the ACI-ASCE Joint Committee 352.

Parameters of the Analytical Model

A hysteresis model based on the concept of accumulated damage was proposed to analyze a multi-degree of freedom reinforced concrete shear wall structure. Originally, the model parameters were to be calibrated from the results of TRG experiments performed by Los Alamos National Laboratory as part of the NRC's Seismic Category I Shear Wall Structure Program; however, the accelerometer data were influenced by the rigid-body motion and base connection effects and the data obtained gave a measure of the combined shake table and test structure properties, and not those of the shear walls alone. This combination made extrapolation of the results in terms of force-displacement relationship difficult (Subia & Wang, 1991). Furthermore, the scaled structure models failed in an abrupt manner as a result of the lack of ductility in their rebar (Farrar, Dove & Baker, 1990). None of the test results showed any yielding at all. Since the proposed nonlinear hysteresis model is based on a complete force-displacement curve and the concept of accumulated damage after yielding, the experiment results were only used in an equivalent stiffness degradation model in the analysis and were unable to be used in the nonlinear hysteresis model. It was then decided to use values of the damage parameters which were calibrated from both scaled and prototype reinforced concrete beam-column joints (Wang & Shah, 1987), including the shear stress effect. In general, it was observed that the damage parameter n was sensitive to the amount of shear reinforcement used and the amount of the shear stress presented; however, parameter n, which determines the rate of damage, appeared to be an insignificant value in the analysis for a lower level of ductility response (about 3) that a structure experiences during an earthquake. It was decided to use n = 1.0 and set C = 0.1 in the analysis. Thus the proposed approach satisfactorily predicted the hysteresis behavior of both scaled (Wang & Shah, 1987) and prototype (Uzumeri, 1977) specimens.

Results have suggested that the skeleton curve or envelope curve for inelastic shear elements representing the shear flexibility of low rise reinforced concrete walls can be defined by a bilinear approximation of the deformation behavior of low aspect ratio walls under monotonically increasing load (Kennedy et al., 1984). The slope of the first segment (K_1) represents the effective elastic shear stiffness of the concrete wall, while the slope of the second segment (K_2) represents the effective stiffness of the reinforced steel after the cracking has occurred as shown in Figure 3. For the skeleton curve, the yield load (F_y) and the corresponding displacement (δ_y) were deduced based on extensive literature review on concrete shear wall studies (Kennedy, Wesley, & Tong). It was decided to approximate a failure based on the maximum base shear stress of 750 psi. and its corresponding failure displacement of 0.6-0.7 % of the height of the structure (Fig. 3). The slope of the post-yield curve was taken as 5 % of the initial calculated effective stiffness (K_1). The yielding displacement was taken as 0.1% of the height of the structure. Then the corresponding failure load (F_f) and yielding load (F_y) were determined.

Structure Modeling

The structure considered in this research was a three-story reinforced concrete shear wall structure (Farrar, Dove & Baker, 1990) as shown in Figure 4. The story height was 420 inches. The walls were 42-inch concrete panels and the building floors and roof were 42-inch concrete slab (Fig. 4). The walls were considered as vertical structure elements and no columns were constructed between floors. The anchorage bars were placed at the connection points between the concrete floor slabs and walls. During the ground motion, the floors and the roof acted as rigid bodies. The inertia forces were transmitted to the foundation through the walls. The structure model used in the analysis was a three-degree-of-freedom lumped system. In the analysis, some of the assumptions were made as follows:

1. Only lateral shear deformations were considered in the concrete shear walls.
2. Lumped mass system was adopted; the lumped masses were concentrated at floor levels.
3. If deformations of the element compared with the size of the cross-section was small, the geometrical nonlinear effects could be ignored.
4. The damping of the structure was Rayleigh damping. The damping matrix was related to the mass matrix and the initial stiffness matrix and remained invariable during vibration.

The structural model parameters were calculated and are given below:

The ultimate failure shear stress:

$$\tau_f = 750 \text{ lb/in}^2$$

The ultimate failure lateral displacement:

$$\delta_f = 0.6\ \%\ H = 2.52 \text{ in}$$

The yielding lateral displacement:

$$\delta_y = 0.1\%\ H = 0.42 \text{ in}$$

The ultimate failure shear force:

$$F_f = A\,\tau_f = 6.88 \times 10^7 \text{ lb}$$

The yielding shear force:

$$F_y = 4.19 \times 10^7 \text{ lb}$$

The maximum elastic shear force:

$$F_e = 30\%\ F_y = 1.26 \times 10^7 \text{ lb}$$

The calculated initial stiffness (Farrar, Bennett, & Dunwoody, 1990) is

$$K_1 = \left(\frac{1}{K_f} + \frac{1}{K_s}\right)^{-1} = 2.56 \times 10^8 \text{ lb/in}$$

where $K_f = \dfrac{12\,EI}{H^3}$ are flexure stiffnesses

and $K_s = \frac{A\,G}{H}$ are shear stiffnesses.

The yielding stiffness is

$K_y = 5\%\ K_1 = 1.28 \times 10^7$ lb/in.

The equivalent stiffness is

$$K_3 = \frac{F_y}{\delta_y} = 9.98 \times 10^7 \text{ lb/in.}$$

The cracking stiffness is

$$K_2 = \frac{F_y - F_e}{\delta_y - \delta_e} = 7.89 \times 10^7 \text{ lb/in.}$$

Masses:

The story weight is

$W = 10.372 \times 10^6$ lb.

The story mass is

$$M = \frac{10.372 \times 10^6 \text{ lb}}{386.4 \text{ in/sec}^2} = 2.68 \times 10^4 \frac{\text{lb}}{\text{in/sec}^2}.$$

These values are also summarized in Figure 3.

Analysis Procedures

Equation of Motion. The equation of motion of a nonlinear multi-degree of freedom shear-wall structure can be written as:

$$[M]\{\ddot{x}\} + [C]\{\dot{x}\} + \{R\} = -[M]\{I\}\ddot{x}_g \tag{6}$$

where [M], [C] are mass and damping matrix, respectively.

$\{\ddot{x}\}$ are relative accelerations.

$\{\dot{x}\}$ are relative velocities.

{R} are restoring forces of the structure.

$\ddot{x}_g$ is ground acceleration.

For linear elastic dynamic analysis, {R} = [K] {x}, [K] is the initial stiffness matrix, and {x} are relative displacements. In this case Equation 6 becomes

$$[M]\{\ddot{x}\} + [C]\{\dot{x}\} + [K]\{x\} = -[M]\{I\}\ddot{x}_g \tag{7}$$

For nonlinear analysis where the stiffness matrix varies with plastic deformations and time t, the tangential or transient stiffness matrix [K(t)] is used, and the equation of motion is usually given in incremental form. For example,

$$[M]\{\Delta\ddot{x}\} + [C]\{\Delta\dot{x}\} + [K(t)]\{\Delta x\} = -[M]\{I\}\Delta\ddot{x}_g \tag{8}$$

where

$\{\Delta\ddot{x}\} = \{\ddot{x}(t+\Delta t)\} - \{\ddot{x}(t)\}$ are increments of relative acceleration from time t to t + Δt,

$\{\Delta\dot{x}\} = \{\dot{x}(t+\Delta t)\} - \{\dot{x}(t)\}$ are increments of relative velocity from time t to t + Δt, and

$\{\Delta x\} = \{x(t+\Delta t)\} - \{x(t)\}$ are the increments of relative displacements from time t to t + Δt.

Model Analysis. Neglecting the effect of damping, the natural frequencies and the vibration mode shapes can be obtained by solving a general eigenvalues problem. For the structures obtained through the eigenvalues equation, the frequencies and mode shapes are obtained as

$\omega_1 = 27.158$ rad/sec

$\omega_2 = 76.095$ rad/sec

$\omega_3 = 109.961$ rad/sec and

$$[\Phi] = [\{\Phi_1\},\{\Phi_2\},\{\Phi_3\}] = \begin{bmatrix} 0.3280 & 0.7370 & 0.5910 \\ 0.5910 & 0.3280 & -0.7370 \\ 0.7370 & -0.5910 & 0.3279 \end{bmatrix}.$$

Determination of Damping Matrix. In this research, damping is modeled as viscous damping. Because of the lack of detail information on damping, the Rayleigh damping method was used in this analysis. The damping matrix is composed by the mass matrix and stiffness matrix:

$$[C] = \alpha[M] + \beta[K], \tag{9}$$

in which α and β are arbitrary proportionality factors that will satisfy the orthogonality condition. With this type of damping matrix it is possible to compute the damping influence coefficients necessary to provide a decoupled system having desired damping ratios in any specified number of modes.

The damping matrix [C] can be transformed into a modal coordinate system by pre- and post- multiplying both sides of Equation 9 by $[\Phi]^T$ and $[\Phi]$, respectively, resulting in

$$\alpha + \beta\,\omega_k^2 = 2\,\zeta_k\,\omega_k \qquad k=1,2,\ldots,n \tag{10}$$

where ζ_k k=1, 2,...,n are called modal damping coefficients. In practice, the first two damping coefficients ζ_1 and ζ_2 were specified and α and β were obtained as

$$\alpha = \frac{2\,\omega_1\omega_2\,(\zeta_1\omega_2 - \zeta_2\omega_1)}{\omega_2^2 - \omega_1^2}$$

$$\beta = \frac{2(\zeta_2\omega_2 - \zeta_1\omega_1)}{\omega_2^2 - \omega_1^2} \tag{11}$$

The other damping coefficients were computed by following equation:

$$\zeta_k = \frac{\alpha+\beta\,\omega_k^2}{2\,\omega_k} \qquad k = 3,4,\dots \dots,n \tag{12}$$

After the value of α and β from Equation 11 were obtained, the damping matrix was determined by Equation 9.

Numerical Solution of Nonlinear Dynamic Equations

Wilson-θ method was used to solve the nonlinear structural dynamic equations. Wilson-θ is an unconditional stable numerical method improved from the linear acceleration method. In the linear acceleration method, it is assumed that the acceleration is varied linearly in the time interval $(t, t+\Delta t)$. Knowing the value of the displacements, velocities and the accelerations at time t, and using the linear acceleration assumption, the displacements, velocities and acceleration at time $t+\Delta t$ can be computed by solving the dynamic equation, Equation 6. The linear acceleration method is a widely used method in structural dynamic analysis. As long as the time increment in each step is less than one-tenth of the natural period of structure, the linear acceleration method is an effective method in solving both linear and nonlinear dynamic equations. But the linear acceleration method is the only a stable method under a certain condition, i.e., $\Delta t < 0{\cdot}55T_m$; where T_m is the minimum natural period of the structure. For structures with a large number of degree-of-freedoms, the minimum natural period is usually very short and the above condition is difficult to satisfy. Based on the linear acceleration method, Wilson-θ method assumes that the linearity of acceleration extends to the outside of the time interval Δt, i.e., in Wilson-θ method, $t = \theta\Delta t$ ($\theta > 1{\cdot}0$) is used to replace the Δt in the linear acceleration method. By inputting the excitation value (force or ground acceleration) at time $t+\tau$, following the procedure of the linear acceleration method, the responses (displacement, velocity and acceleration) at $t+\tau$ are obtained first. Then the responses at time $t+\Delta t$ are obtained by linear interpolation.

If $\theta = 1$, Wilson-θ method is equivalent to the linear acceleration method. When $\theta > 1{\cdot}37$ the stability of Wilson-θ method is unconditional.

RESULTS

A prototype three-story auxiliary building subjected to both Iran and El Centro earthquakes was analyzed by using the proposed hysteresis model and equivalent stiffness degradation and linear elastic models. The maximum relative displacement, permanent displacement, maximum shear stresses and maximum acceleration excited for each story were compared for both cumulative and one shot inputs. Recent experimental results indicate that the stiffness degradation depended only on the maximum acceleration input or maximum displacement attained during the excitation (Farrar, Dove, & Baker, 1990). No comparable accumulation of damage was observed between the specimen subjected to consecutive increasing amplitude excitations and a one shot excitation at the same peak ground acceleration. This finding is congruent with other research results in the same area (Wang & Shah, 1987). Since none of the experimental results showed any yielding, the specimens were presumably subjected to linear excitations. Accumulation of damage usually happened after yielding. This occurrence is one of the basic assumptions of a damage-based hysteresis model like the one proposed in this study. In-structure floor absolute acceleration response spectra for each story were also compared for different models.

Iran earthquake time history is shown in Figure 5. The frequency content is concentrated between 2 to 8 Hz. The prototype three-story auxiliary building elevation and dimensions are shown in Figure 4. A hysteretic force-deflection model representing the measured behavior shown in Figure 3 was used in the analysis. Structural hysteresis properties are also summarized in Figure 3. A multi-degree nonlinear analysis program for reinforced concrete shear wall structures was then developed. First, the structure was subjected to a Tabas Iran earthquake with a maximum PGA level of 0.8 g. The structure was also subjected to a consecutive series of increasing PGA levels of the same earthquake. The objective of this analysis was to understand the accumulation of damage effects on the structural response. For this level of excitation (ductility is about 2), it seems that cumulative loadings have no effect on the responses as shown in Table 1. The structure needs a PGA level of 1.8 g in order to achieve the maximum shear stress of 720 psi, as shown in Table 1, which is very close to the failure shear stress of 750 psi. Yielding in the shear wall is slight and only limited to the first floor for 0.8 g PGA level as shown in Figure 6. Hysteresis diagrams corresponding with PGA level of 1.2g are shown in Figure 7. Stiffness and strength degradation are dramatic, which is typical behavior when reinforced concrete structures are subjected to high ductility loading (more than 6).

In-floor absolute acceleration response spectra (SRS) were usually used to qualify the safety equipment installed in each floor. Absolute acceleration response spectra due to Iran earthquake with PGA of up to 0·8g are shown in Figure 8. When compared to the in-structure floor SRS using linear elastic analysis as shown in Figure 9, the differences were apparent. In the linear analysis, we used the original calculated stiffness K_1. In general, the result using linear analysis showed a higher value of absolute acceleration than the result using the nonlinear hysteresis model; however, the hysteretic model predicted a wider band of frequency distribution, especially in the high frequency range centered at second mode frequency with significant values of acceleration response up to 5g. These high frequency ranges result in significant safety implications for equipment placed within their parameters on certain floors. The high acceleration response in high frequency components was more pronounced when the structure was subjected to high PGA levels.

CONCLUSION

A Hysteresis model based on the concept of accumulation of damage has been developed to account for both stiffness and strength reductions. A multi-degree-of-freedom shear wall structure was analyzed. The results were compared with linear analysis without considering the stiffness degradation and an equivalent linear stiffness degradation analysis in terms of maximum displacement responses, permanent displacement, maximum shear stress, and floor response spectra. The results indicate significant differences in responses between a hysteresis model and equivalent linear stiffness degradation and linear models. Equivalent stiffness degradation models tend to soften the structure and underpredict the in-structure floor acceleration responses. On the other hand, the hysteresis model predicted a wider band of frequency power distributions and a larger magnitude of absolute floor acceleration responses centered both at the first and second modes. This prediction was especially accurate when the structure was subjected to high input ground acceleration. Ductility and hysteresis energy played major roles in determining the structure responses. The floor acceleration responses due to the effects of higher modes were significant. Neither linear nor equivalent stiffness degradation models can reflect these effects. Therefore, it is recommended to use the hysteresis model to determine the in-structure floor responses for equipment design and reliability analysis.

REFERENCES

Bennett, J.G., et al. *The Seismic Category I Structures Program: Results of FY 1985*, Los Alamos National Laboratory Report, NUREG/ CR-4998, 1987.

Bennett, J.G. et al. *Simulated Seismic Tests on 1/42 and 1/14 scale Category I, Auxiliary Building*, Los Alamos National Laboratory Report, NUREG/ CR-4987, 1987.

Dove, R.C. & Bennett, J.G. *Scale Modeling of Reinforced Concrete Category I Structures Subjected to Seismic Loading*. Los Alamos National Laboratories Report, NUREG / CR-4474, 1986.

Dove, R.C. et al. *Seismic Category I Structures Program Final Report, FY 1983-84*, Los Alamos National Laboratory Report, NUREG / CR-4924, 1987.

Endebrock, E.G. et al. *Analysis and Tests of Small Scale Shear Walls FY-82 Final Report*, Los Alamos National Laboratories Report, Nureg/ CR-4274, 1985.

Farrar, C.R., Bennett, J.G., and Dunwoody, W.E. *The Seismic Category I Structure Program Results for FY 1987*. Los Alamos National Laboratory Report, NUREG/ CR-5369, 1990.

Farrar, C.R., Dove, R.C. and Baker W.E. *Static and Simulated Seismic Testing of TRG-12 Through 16 Results*. Los Alamos National Laboratory Report, 1990.

Kennedy, R.P. et al. *Engineering Characterization of Ground Motion. Task I: Effects of Characteristics of Free-Field Motion on Structural Response*, NUREG/CR-3805, Vol. 1, 1984.

Kennedy, R.P., Wesley, D.A. and Tong, W.H. *Probabilistic Evaluation of The Diablo Canyon Turbine Building Seismic Capacity Using Nonlinear Time History Analysis*, NTS Engineering Report, 1643.01.

Subia, S. & Wang, M.L. *Computation of Displacement Time Histories by Direct Numerical Integration of Acceleration Data*, Submitted to The International Journal of Analytical and Experimental Modal Analysis, April, 1991.

Townsend, W.H. and Hanson R.D. *Reinforced Concrete Connection Hysteresis Loops*, ACI Publication SP 53-13, 1977. 351-370.

Uzumeri, S.M. *Strength and Ductility of Cast-in-place Beam-Column Joints*, ACI Publication SP 53-13, 1977. 283-305.

Wang M.L. and Shah S.P. *Reinforced Concrete Hysteresis Model Based on the Damage Concept*, Journal of Earthquake Engineering and Structural Dynamics, 15. 1987. 993-1003.

Wang, M.L. and Shah, S.P. *Model Reinforced Concrete Beam-Column Joint Behavior Subjected to Cyclic Loading at Two Rates*, Journal of Materials and Structures, 20. 1987. 85-95.

TABLE 1. COMPARISON OF RESPONSES AT VARIOUS PGA LEVELS (TABAS, IRAN EARTHQUAKE)

	Floor	Excitation acceleration		
		0·8 g	(0·2 – 0·6 + 0·8)g	1·8 g
Max. disp. (in)	1	0·74	0·796	2·53
	2	0·36	0·353	0·505
	3	0·206	0·2	0·274
Perm. disp. (in)	1	0·14	0·18	0·8
	2	0	0	0·1
	3	0	0	0
Max. accel. (g)	1	1·66	1·63	2·38
	2	1·8	1·8	2·27
	3	2·05	2·0	2·83
Max. Force. ($lb*10^7$)	1	4·44	4·36	6·6
	2	3·6	3·52	4·2
	3	2·06	1·99	2·74
Max. shear stress (lb/in^2)	1	484	475	720
	2	392	384	458
	3	224	217	299

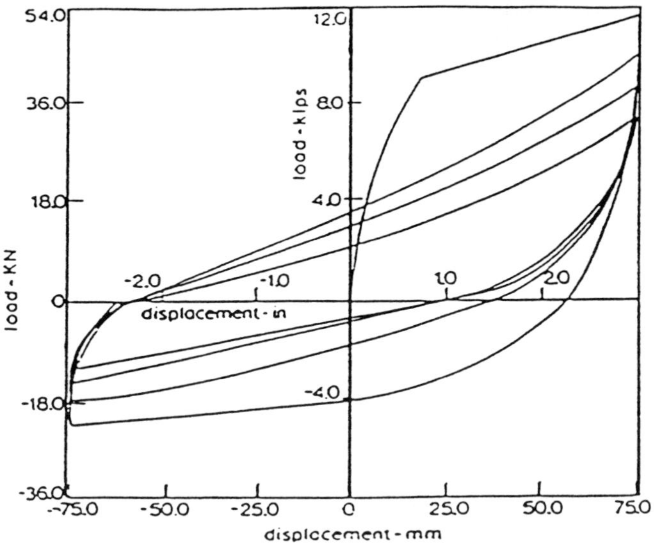

FIG. 1. HYSTERESIS DIAGRAM FOR SPECIMEN 10 OF REFERENCE.

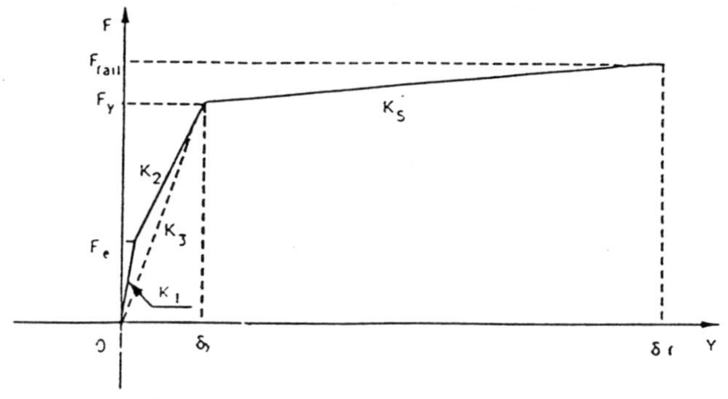

$M = 2{\cdot}68 \times 10^4$ lb s²/in $H = 420$ in $A = 2 \times 45864 = 91728$ in²

$\tau_f = 750$ lb/in² $\delta_f = 0{\cdot}6\%H = 2{\cdot}5$ in $\delta_y = 0{\cdot}1\%H = 0{\cdot}42$ in

$F_f = A \times \tau = 6{\cdot}88 \times 10^7$ lb $F_y = 4{\cdot}19 \times 10^7$ lb $F_e = 0{\cdot}3F = 1{\cdot}26 \times 10^7$ lb

$\tau_y = F_y/A = 453$ lb/in² $K_1 = 2{\cdot}56 \times 10^8$ lb/in $K_2 = 7{\cdot}89 \times 10^7$ lb/in $= 30{\cdot}9\% K_1$

$K_3 = 9{\cdot}98 \times 10^7$ lb/in $= 39\% K$ $K_s = 5\% K_1 = 12{\cdot}8\% K_3 = 1{\cdot}28 \times 10^7$ lb/in

FIG. 3. PROPOSED SKELETON CURVE FOR SHEAR WALL STRUCTURE AS SHOWN IN FIG. 7

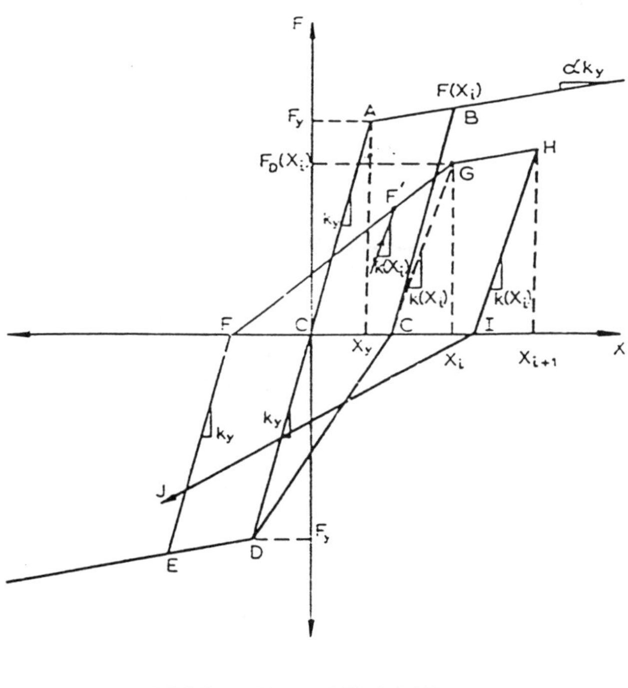

FIG. 2. PROPOSED MODEL LOAD-DISPLACEMENT CURVE.

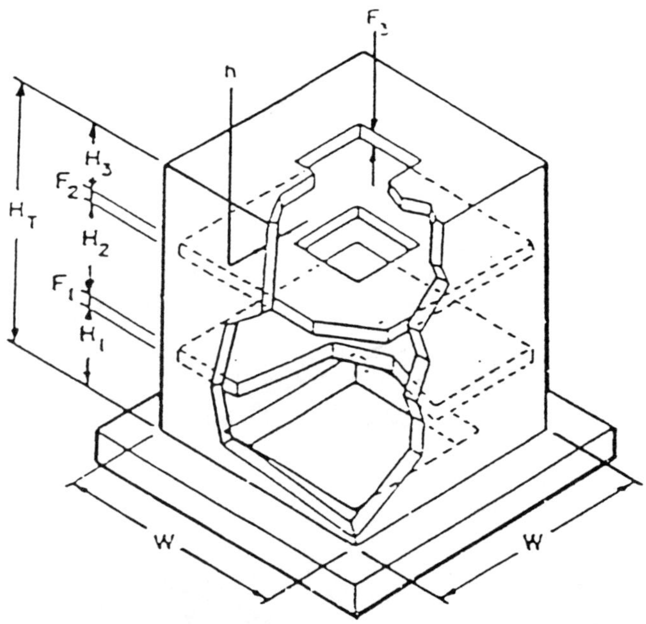

	h, F₁, F₂, F₃	W	H₁, H₂, H₃	Wt/STORY* (lb)
1/42 SCALE	1 in.	26 in	10 in	140
1/14 SCALE	3 in.	78 in.	30 in.	3780
PROTOTYPE	42 in.	1092 in.	420 in.	10,372,000

*BASE NOT INCLUDED

FIG. 4. SHEAR WALL STRUCTURE USED IN THE ANALYSIS.

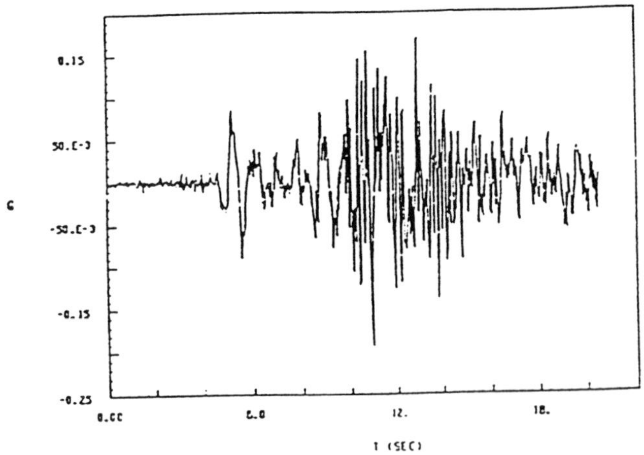

FIG. 5. TIME HISTORY (TABAS, IRAN EARTHQUAKE)

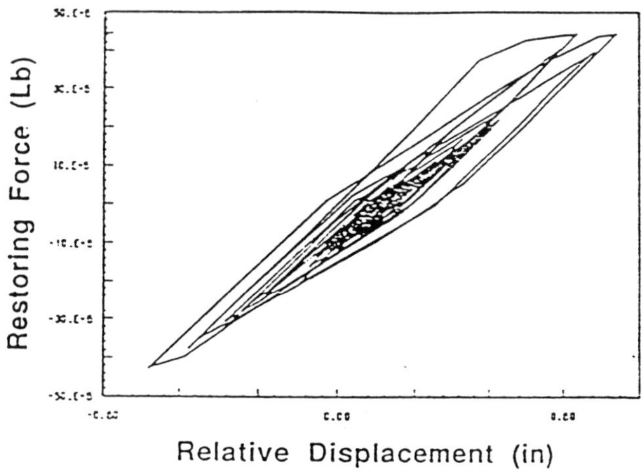

FIG. 6. FIRST FLOOR HYSTERESIS DIAGRAM USING WANG'S MODEL SUBJECTED TO IRAN EARTHQUAKE AT PGA = 0·8g.

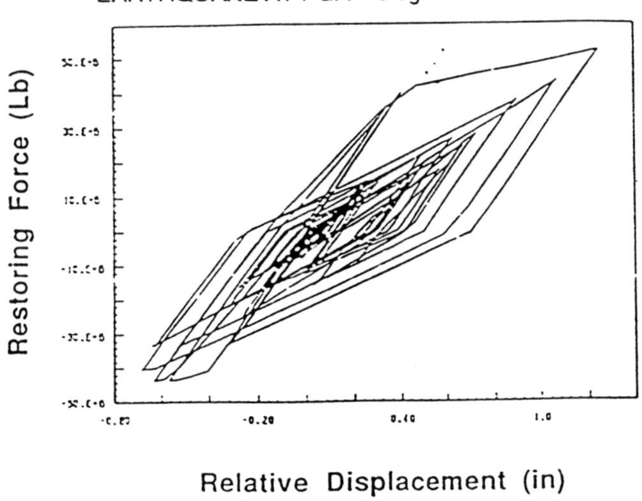

FIG. 7. FIRST FLOOR HYSTERESIS DIAGRAM USING WANG'S MODEL SUBJECTED TO IRAN EARTHQUAKE AT PGA = 1·2g.

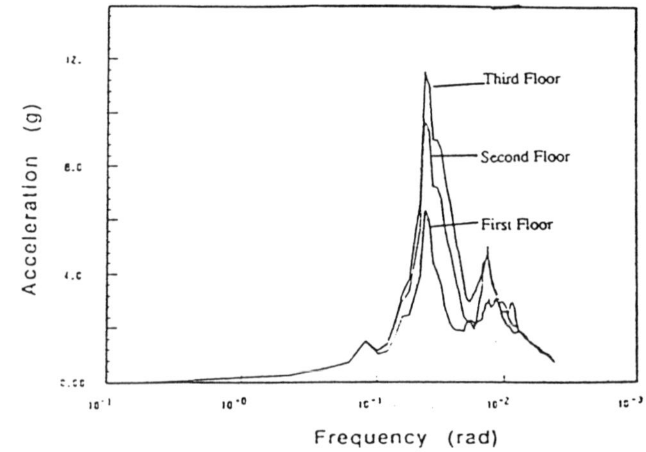

FIG. 8. ABSOLUTE ACCELERATION FLOOR RESPONSE SPECTRA (SRS). (WANG MODEL, IRAN EARTHQUAKE, PGA = 0·8g).

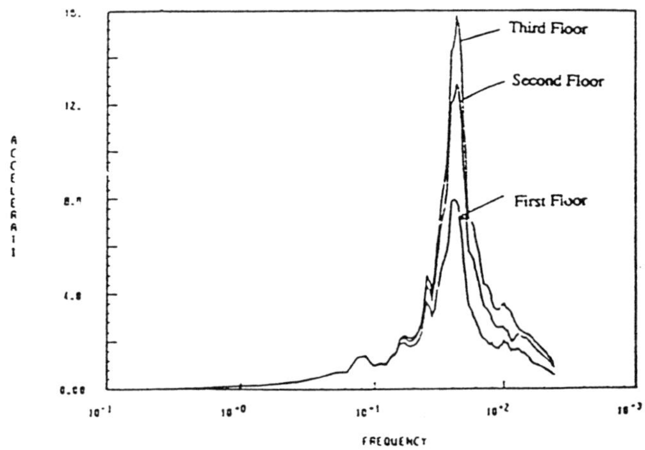

FIG. 9. ABSOLUTE ACCELERATION FLOOR RESPONSE SPECTRA (SRS). (LINEAR ELASTIC, IRAN EARTHQUAKE PGA = 0·8g).

AUTHOR INDEX

Vibrations of Mechanical Systems and the History of Mechanical Design

Book Number: G00824